DIAGNOSTIC ULTRASOUND AND ANIMAL REPRODUCTION

Current Topics in Veterinary Medicine and Animal Science

Volume 51

For a list of titles in this series see final page of this volume.

Diagnostic Ultrasound and Animal Reproduction

Edited by

M. A. M. TAVERNE

and

A. H. WILLEMSE

Department of Herd Health and Reproduction,
Veterinary Faculty, University of Utrecht, The Netherlands

SPRINGER-SCIENCE+BUSINESS MEDIA, B.V.

Library of Congress Cataloging in Publication Data

Diagnostic ultrasound and animal reproduction / edited by M.A.M.
Taverne & A.H. Willemse.
p. cm. -- (Current topics in veterinary medicine and animal
science ; v. 51)
Includes index.
ISBN 978-90-481-4053-4 ISBN 978-94-017-1249-1 (eBook)
DOI 10.1007/978-94-017-1249-1
1. Veterinary obstetrics. 2. Veterinary ultrasonography.
I. Taverne, M. A. M. (Marcel A. M.) II. Willemse, A. H.
III. Series.
SF887.D53 1989
636.089'8207543--dc20 89-15429

ISBN 978-90-481-4053-4

TABLE OF CONTENTS

PREFACE

The use of ultrasonic imaging techniques for diagnostic purposes in veterinary medicine and animal science has lagged far behind their use in human medicine. In the area of domestic animal reproduction, diagnostic ultrasonography has a relatively short history. Reports on B-mode scanning first appeared as late as 1969, when Stouffer and co-workers used it for counting foetal numbers in late gestation sheep. After Lindahl had re-evaluated the potential of two-dimensional ultrasonography for pregnancy diagnosis in sheep in 1976, results of the first large scale field-trials by Fowler and Wilkins on predicting foetal numbers during the first half of gestation in sheep were published as recently as 1980. In this year also the first paper, by Palmer and co-workers, on ultrasonic scanning of the uterus and ovaries in the mare appeared. The improvement of diagnostics which was achieved by the application of this technique is illustrated by its present routine use in clinical equine practice and sheep breeding. Besides offering an accurate early pregnancy diagnosis in species such as the horse, cow, sheep, goat, pig and dog, ultrasonography enables the visualization of ovarian and uterine structures for the identification of both physiological and pathological conditions. In this way it replaces or supplements the more common diagnostic techniques used by the general practitioner so far. Ultrasonography has also opened several new lines of research for clinicians and research workers. However, many of these potential users are not yet familiar with the basic principles of ultrasound imaging techniques and the possibilities for application under experimental and farm conditions. This book provides such basic information.

After an exploration of the most relevant technical principles of diagnostic ultrasonography, several descriptive papers document the ultrasonic findings which can be expected on scans of the uterus and ovaries of mares and cows. Other chapters focus on the accuracy of ultrasound use for early pregnancy diagnosis (cows, goats, pigs and dogs), foetal number determination (sheep, goat) or estimation of fetal age (sheep, cow). It should be kept in mind that these quantitative date are essential for estimating the profitability of the use of ultrasonography. Finally, a short survey is given of the possible applications in other areas of veterinary medicine and current applications of ultrasonography in human reproduction are reviewed.

To our knowledge, this collection of papers is the first of its kind and is intended for veterinary and animal science students, general practitioners and university teachers involved in introductory scanning courses.

Objective criticism and suggestions will help us improve our efforts to convince others that ultrasonography should be an integral part of the standard techniques used in clinical practice and animal breeding management.

M.A.M. Taverne
A.H. Willemse
April 1989

PREFACE

The use of ultrasonic imaging techniques for diagnostic purposes in veterinary medicine and animal science has lagged far behind their use in human medicine. In the area of domestic animal reproduction, diagnostic ultrasonography has a relatively short history. Reports on B-mode scanning first appeared as late as 1969, when Stouffer and co-workers used it for counting foetal numbers in late gestation sheep. [illegible] re-evaluated the potential of two-dimensional ultrasonography for pregnancy diagnosis in sheep in 1976. Results of the first large scale field trials by Fowler and Wilkins on predicting foetal numbers during the first half of gestation in sheep were published as recently as 1980. In this year also the first papers, by Palmer and co-workers, on ultrasonic scanning of the uterus and ovaries in the mare appeared. The improvement of diagnostics which was achieved by the application of this technique is illustrated by its present [illegible]

[illegible] and [illegible], however, many of these potential users are not yet familiar with the basic principles of ultrasound imaging techniques and the possibilities for application under experimental and farm conditions. This book provides such basic information.

[illegible]

[illegible] and current applications of ultrasonography in human reproduction are reviewed.

To our knowledge, this collection of papers is the first of its kind and is intended for veterinary and animal science students, general practitioners and university [illegible] involved in introductory scanning courses. Objective criticism and suggestions will help us improve our efforts to convince others that ultrasonography should be an integral part of the standard techniques used in clinical practice and animal breeding management.

M.A.M. Taverne
A.H. Willemse
April 1989

TECHNICAL PRINCIPLES OF ULTRASOUND

C.M. Ligtvoet, N. Bom and W.J. Gussenhoven
Academic Hospital Utrecht, Erasmus University Rotterdam and the Interuniversity Cardiology Institute, The Netherlands

INTRODUCTION

Diagnostic ultrasound systems have become increasingly sophisticated and instruments now exist which can combine more than one diagnostic parameter that can be obtained with ultrasound. Four of these diagnostic parameters obtainable with ultrasound are shown in Fig. 1.

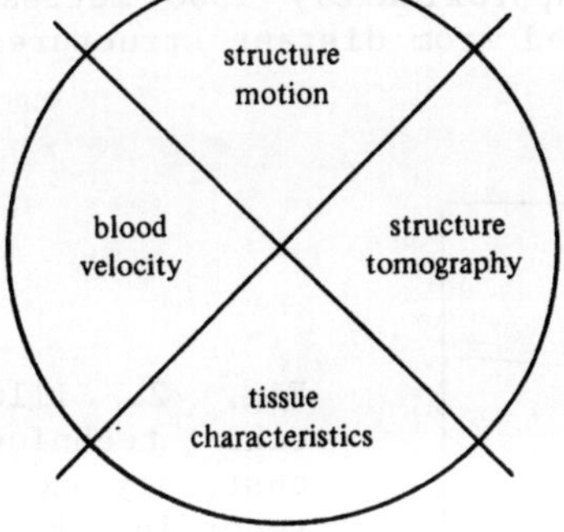

Fig. 1. Potential diagnostic information. With the use of ultrasound, diagnostic information can be obtained in the indicated categories.

With structure tomography it is possible to reconstruct the cross-sectional anatomy of an organ; structure motion allows observation of moving objects such as the fetus or cardiac valves; blood velocity can be measured with Doppler and approaches towards tissue characterization with ultrasound are also available. Currently the most important areas in diagnostic ultrasound are structure tomography and structure motion observation.

It is well-known that cross-sectional echo images can be constructed if sufficient echo dots are available. In order to achieve this result, modern real-time systems rapidly steer their soundbeam through the organ allowing observation of moving structures. This development took many years.

As early as 1952 Wild and Reed publised their first results of "A two-dimensional picture such as would be obtained by adding up the information from a series of needle biopsies taken in one plane". They described the first principle of a mechanical sector scan. The idea to compose cross-sectional images by aiming the beam from more than one angle dates back to 1958 and was introduced by Donald and co-workers. Important cardiological breakthroughs in ultrasound were initiated in 1954 by Edler and Hertz. Today it is impossible to conceive of

M. M. Taverne and A. H. Willemse (eds.), Diagnostic Ultrasound and Animal Reproduction, 1–9.

diagnostic units without ultrasound equipment to complement the diagnostic apparatus. Ultrasound investigation is particularly suitable for study of geometry and motion of soft tissues. Important areas include ophthalmology, cardiology, abdominal studies, obstetrics and gynaecology. In this paper the principle of ultrasound technique, in particular the real-time imaging systems, will be presented. A comparison between the earlier compound scanning methods and the modern real-time scanners will be made.

THE ECHO PRINCIPLE

Sound waves represent a pressure disturbance that is propagated with a given sound velocity through a medium such as water or tissue. The sound velocity in blood or water is approximately 1500 metres per second. Short sound pulses may be reflected from distant structures such as the ocean bed.

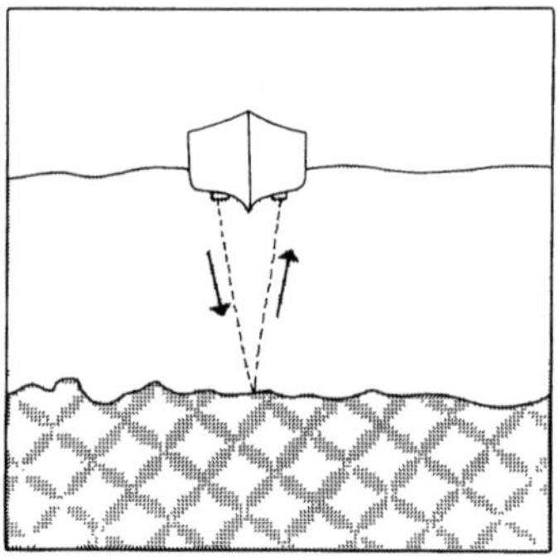

Fig. 2. Illustration of echo techniques by the customary example of depth recording by echo travel time.

Figure 2 shows a well-known application of measuring the depth of water beneath a ship. With known sound velocity the time elapsed between transmission of a sound pulse and reception of the echo can be converted to a distance or depth measurement. The longer the echo waiting time the greater the depth.

ULTRASONIC TRANSDUCER

Sound waves used in clinical ultrasound have a frequency between 1 and 10 MHz, depending on the application, and can be generated with piëzo electric material.

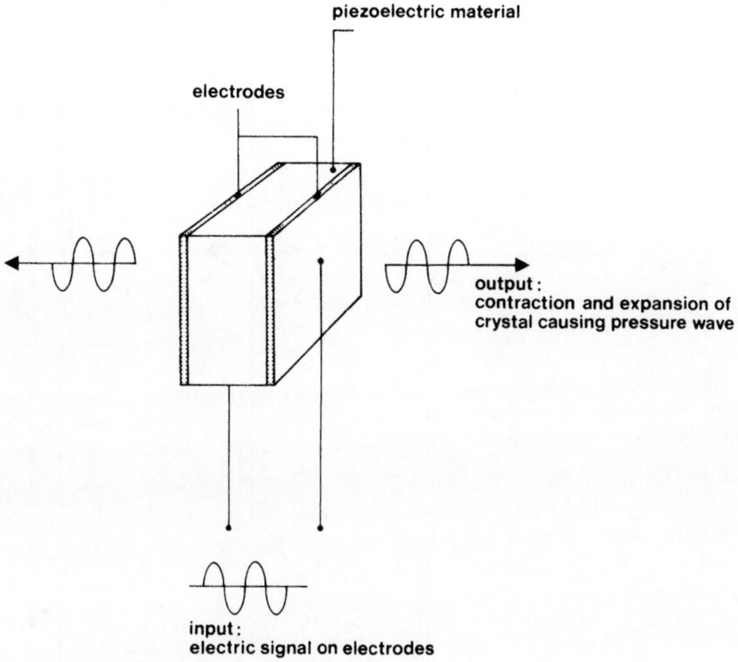

Fig. 3. The Piëzo-electric effect.

Figure 3 shows a schema of the principle. A thin piëzo electric crystal is covered, on both sides with electrodes. When an electric signal with the appropiate frequency is applied to the electrodes, the crystal becomes thinner and thicker in relation to the same frequency. In this way, sound waves are generated at the surface of the crystal, which are then propagated in two directions. With damping material, one of these waves is suppressed and the other wave penetrates the tissue in front of the crystal. The damping material, the electrodes, the piëzo electric crystal together with the housing form the ultrasonic transducer.

When a reflected sound wave reaches the transducer, an electric signal is measured over the electrodes. In this way ultrasound waves can be generated and detected.

REFLECTIONS

Tissue may be acoustically defined by its "acoustic impedance", which is the product of the sound velocity and the density of the tissue. The reflectivity at boundaries between two media depends on their difference in acoustic impedence. Highly reflecting boundaries are, for instance, interfaces between water and air, the epicard/lung interface, air bubbles in the colon and calcified structures.

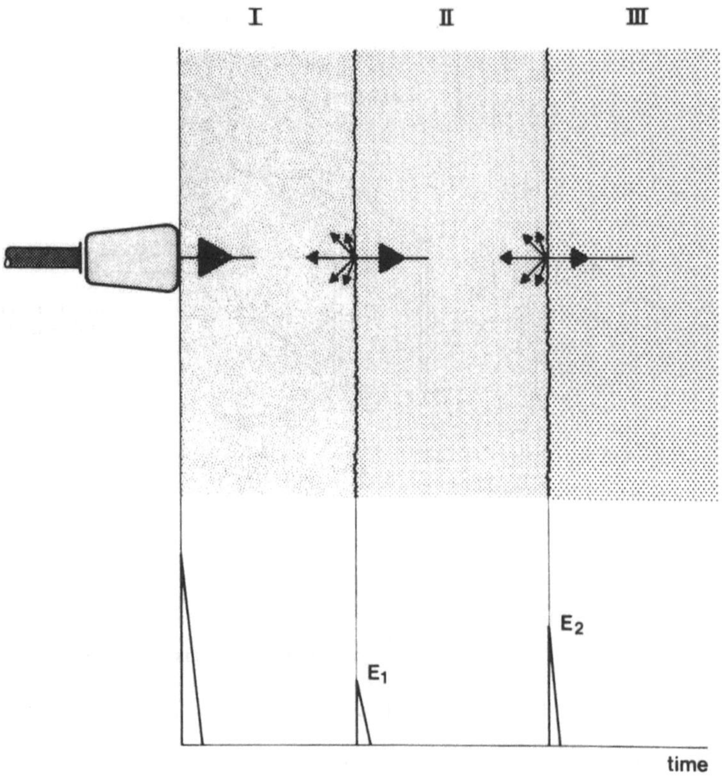

Fig. 4. Reflection of ultrasound at the boundaries of tissue with different acoustic impedence.

In Figure 4 the reflection at boundaries between media I, II and III is diagrammatically presented. From each transmitted pulse only part of the acoustic energy is reflected back to the transducer as an echo. Corresponding with the depth of the reflecting boundary the echo will be displayed on the screen.

DISPLAY TECHNIQUES

As shown in Figure 4 echoes may be displayed in the A-mode or amplitude mode. Apparently the reflectivity at boundary II/III causes echo E2 to be larger than echo E1. Another display fashion is the B-mode, or brightness mode. Here the echo amplitude is converted to brightness of the echo dot. The A-mode is used for very precise measurements, for example in ophthalmology. The B-mode is the basis of all two-dimensional images.

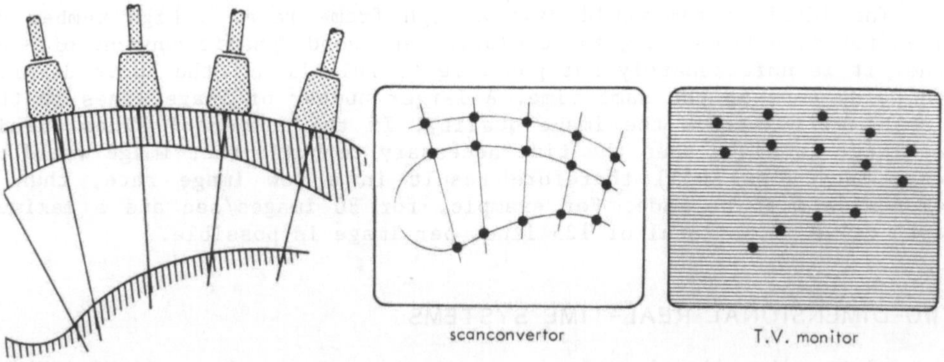

Fig. 5. Composition of an ultrasound image by repositioning the transducer in respect to the investigated structure. The information is written into a digital scanconverter and displayed on a television screen.

In Figure 5 the principle of cross-sectional scanning and the construction of a cross-sectional ultrasonic tomogram is diagrammatically shown. Repeated repositioning of the transducer will cause the ultrasound beam to penetrate the organ along many lines in a well-defined cross-sectional plane. Echoes will be created, particulary at borderlines between blood and tissue, fat and tissue, etc. Corresponding with each transducer position the echo dots can be stored in a scanconverter memory on a line corresponding with the transducer beam at the moment the echoes were oreated. Within the scanconverter all the integrated echo information will form a cross-sectional image of the organ. The image will present a still-frame of the studied structure. The contents of the scanconverter memory may subsequently be displayed an a television monitor.

In early systems the manual compound scanning technique was used. With compound scanning each tissue point is insonified by the ultrasound beam from more than one angle. The manual motion of the transducer is, of course, time-consuming and therefore it is not possible to observe moving organs. Effort have been made to develop systems whereby the ultrasound beam can be rapidly swept through the entire cross-section. This has resulted in electronic as well as mechanical systems whereby a high frame rate becomes possible, the real-time systems.

Characteristics of real-time systems include:

a. Possibility to study structures in motion. This is important for cardiac and fetal studies.

b. During observation the scan plane can be easily selected since the echogram appears instantaneously on the display.

When frame repetition rate is higher than 15 frames/sec the eye will observe the frame sequence as a continuum.

The ideal system would have a high frame rate, a high number of lines for each frame, a good resolution and good dynamic content of each echo. It is unfortunately not possible to fulfill all the above described parameters at the same time. A larger number of image lines in the display will improve the image quality. If the number of lines within one image is high, then the time necessary to build that image would be long. Many lines will therefore result in a low image rate, thus a compromise must be made. For example, for 30 images/sec and a maximum depth of 20 cm, a total of 125 lines per image is possible.

TWO-DIMENSIONAL REAL-TIME SYSTEMS

Several two-dimensional real-time systems are currently available. These may be subdivided into mechanical versus electronic beam steering on sector versus rectangcular image format. The principle of all these systems is that an acoustic beam scans the cross-sectional plane at a high rate so as to yield instantaneous information about two-dimensional geometrical structures. The four main principles of ultrasonic systems are shown schematically in Figure 6.

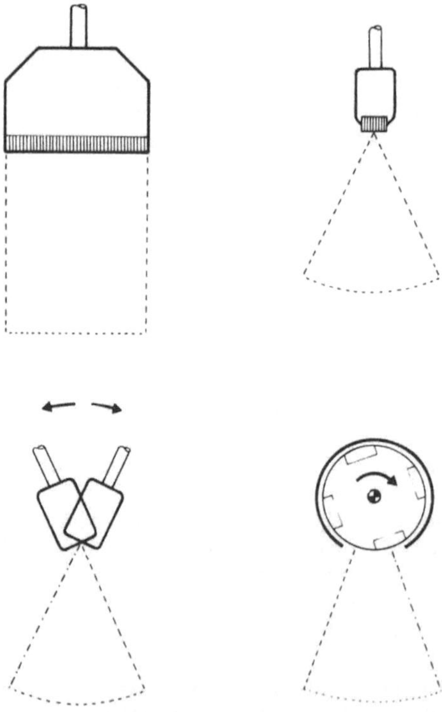

Fig. 6. Schematic diagram of commonly used two-dimensional real-time imaging systems. Left upper: linear array scanner. Right upper: phased array scanner. Left bottom: pivoting mechanical sector scanner. Right bottom: spinning wheel mechanical sector scanner.

1. The linear array method, where a number of piëzo electric crystals have been positioned in a row. A subgroup of crystals is used to form a sound beam. Adjacent and overlapping subgroups produce parallel sound beams. Rapid scanning results in high frame rate. The characteristics of this system are a rectangular image format produced with no mechanical motion of the transducer head, but requiring a relatively large contact area with the patient.
2. Phased array sector scanners also have a stationary transducer with multiple small crystals in a row. All elements are activated simultaneously to produce a single sound beam. Rapid sectorial beam steering is then introduced by electronic time delay processing. The transducer is relatively small. The characteristics are a sectorial image format, with no mechanical motion of the transducer and with excellent probe manoeuvrability and a small contact area.
3. The mechanical sector scanner where pivoting motion is applied to a single transducer. Commonly a small motor and the transducer are combined in a single housing. As an alternative to mechanical pivoting systems, either vibration, magnetic deflection or rotation mechanisms are used to cause sectorial deflection of the transducer. The characteristics are a sectorial image format, sometimes with slight mechanical vibration of the transducer assembly and the possibility of point entry through a small area.
4. The spinning wheel method. With three or four transducers on a rotating wheel, it is possible to create sectorial beam deflection at a high rate of repetition. Each transducer is active when it passes an acoustic window in the housing. The characteristics are a sectorial image format with mechanically driven rotating transducers and possibility of point entry. The total transducer assembly is usually of larger format compared to other systems.

RESOLUTION LIMITATIONS

The resolution of an ultrasound system can be defined as the system"s ability to distinguish closely related structures. In general, resolution is divided into axial resolution and lateral resolution. The axial resolution is the power of the system to separate closely lying structures along the propagation direction of the sound wave: the beam axis. The lateral resolution is the power of the system to separate closely lying structures in a plane perpendicular to the beam axis.

Axial resolution is determined by the length of the ultrasound pulse generated by the transducer. This can be achieved by choosing a high ultrasound frequency. However, effective penetration depth decreases with frequency. Therefore, a compromise must be made between penetration depth and axial resolution.

The major contribution to image quality and system resolution is the lateral resolution. The lateral resolution depends on the shape of the ultrasound field, produced by the transducer. This sound beam is not needle shaped. Near the transducer the sound beam has the same width as the transducer. After a certain depth the sound field diverges. Both parts of the sound field are within the region of diagnostic interest.

In principle all reflectors positioned within the sound field produce an echo, which will be displayed as if it originated on the main axis. This leads to ambiguity in ultrasound images. In two-dimensional systems a number of sound beams is used to compose a two-dimensional image. Due to the limited lateral resolution adjacent sound beams overlap. For example a point reflector in front of the transducer is not only "seen" by one beam, but also by adjacent beams. Consequently, on the display the point reflector will be displayed several times on lines which correspond with these beams. Instead of an echo point, a "banana" shaped echo appears. This is shown in Figure 7.

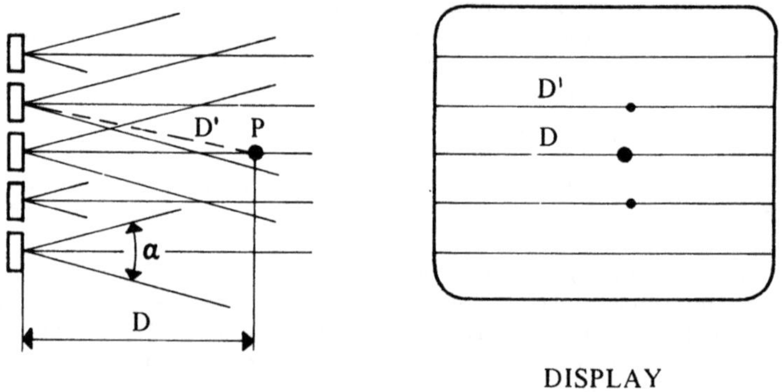

Fig. 7. Schematic representation of lateral distortion in a linear array system. The point reflector P is "seen" as a banana shaped curve due to the diverging sound beam of the individual elements.

ATTENUATION

Ultrasound waves are heavily attenuated in tissue. As a result echoes from a great depth will be weak and additional amplification, depending on depth, is required. This is implemented with the time gain amplification available on all apparatus. Moreover, attenuation is frequency dependent. High ultrasound frequencies suffer heavy attenuation. Therefore, high frequencies such as 10 MHz can only be used for eye studies where penetration is limited to a few centimetres. Bone attenuates excessively, thus neurologists must operate in the 1 MHz range in order to penetrate the skull.

REVERBERATION

Strongly reflecting interfaces may cause echoes to oscillate between two such interfaces before the echo is received by the transducer. Under these conditions, multiple echoes are formed. The echo travel time of the additional echoes is then increased and mirror images may appear at deeper positions. This is further illustrated in Figure 8.

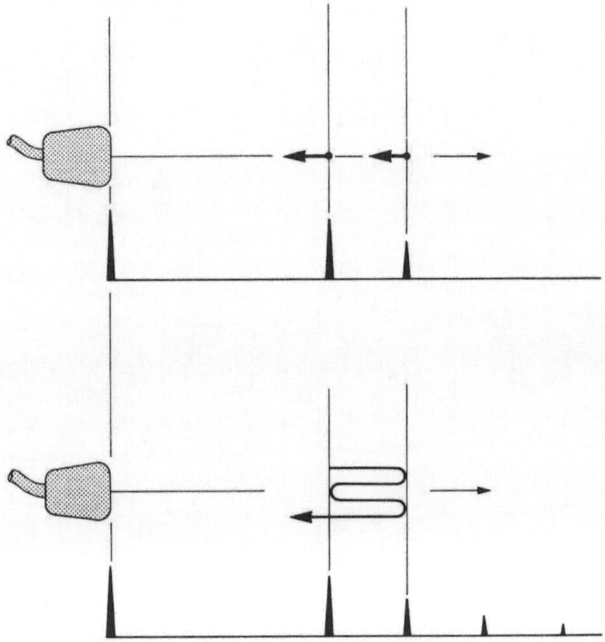

Fig. 8. The mechanism of reverberation between two strongly reflecting boundaries is illustrated. Additional echoes appear on the display.

CONCLUSION

Ultrasound has become very important for non-invasive observation of soft structures in real-time. It is an interactive diagnostic technique, which requires much experience to find the right view for proper diagnosis. Knowledge of the background of some physical principles will help avoid errors in the practical routine use of the method.

RECOMMENDED LITERATURE:

Peter N.T. Wells (1977) "Biomedical Ultrasonics", Academic Press, ISBN: 0-12-742940-9.

Kenneth J.W. Taylor (1978) "Atlas of grey scale ultrasonography", Churchill and Livingstone, ISBN: 0-443-08001-1.

THE USE OF ULTRASONOGRAPHY IN THE REPRODUCTIVE MANAGEMENT OF MARES IN THE FIELD.

P. Fontijne[*] *and C. Hennis*[**]
* *Department of Herd Health and Reproduction*
State University of Utrecht
Yalelaan 7, 3584 CL Utrecht, The Netherlands
** *Veterinary practitioner, Garijp, The Netherlands*

INTRODUCTION

Since the early eighties ultrasound equipment has been available for veterinary use. After a tentative start an explosive growth in the application of this technique in veterinary practice followed. Besides being of use within other disciplines, ultrasound has proved to be a valuable asset in the reproductive management of mares.

The application of this technique for physiological and pathological studies has accelerated reproduction research. Apart from this scientific importance it also offers the practitioner new diagnostic possibilities. Instead of having to form a virtual mental image of the uterus and ovaries during rectal palpation, it is now possible to obtain a real image using a simple procedure.

In addition to its diagnostic value in a practice, a scanner may be used as a teaching aid in rectal exploration or for personal training.

The practical application of real-time ultrasound in the reproductive management of mares will be discussed in the following section. In a number of cases the scanner will provide extra information, but there are instances where the information obtained using ultrasound will be more of less the same as that obtained by manual rectal palpation. What then is the additional diagnostic value offered by the use of a scanner, compared to the more conservative investigative methods?

ULTRASONOGRAPHY YIELDS MORE INFORMATION THAN MANUAL RECTAL PALPATION.

The ovary

Ultrasound is useful for visualizing the funtional structures of the ovary. Pathological changes in and around the ovary may also be diagnosed. The 5 mHz transducer is best suited for the evaluation of ovarian activity because of its good resolution: follicles as small as 2-4 mm may be discerned (Ginther and Pierson, 1984a).

Because of its typical location, the presence of the corpus luteum in the ovary cannot be determined by manual means. However, using a scanner the corpus luteum can be visualized in about 88% of the cases (Ginther and Pierson, 1984b).

M. M. Taverne and A. H. Willemse (eds.), Diagnostic Ultrasound and Animal Reproduction, 11–19.

First ovulation in the spring.

In a number of cases it may be necessary to ascertain early in the mating season whether a mare is cyclic or not. If exclusive use is made of manual examination, it may be difficult to distinguish between a mare in pro-oestrus and one that has not yet ovulated. Both may show a weak signs of oestrus, and the diminshing corpus luteum of the cyclic mare cannot be palpated. Follicular activity is not a deciding factor either: in both cases large follicles, which again diminish in number a few days before ovulation, may be present. This period may be extended before the first ovulation in the season. When examined again, ten days later, the cyclic mare will have ovulated. Using the ultrasound scanner it is nearly always possible to ascertain whether a mare is cyclic or not after a single examination. Should no corpus luteum be found, and the largest follicle is smaller than 20 mm, the mare is anovulatory.

No signs of oestrus during the oestrous season.

When a mare fails to show signs of oestrus during the mating season, manual rectal examination can only supply information about follicular activity; a corpus luteum, after all, is not palpable. The ovaries are of a standard size while the uterus may have a slight tonus, which leads one to suspect the presence of a corpus luteum. In such cases treatment with prostaglandins may induce oestrus, or else a therapy with progestativa is the obvious solution.

In the majority of cases a scanner may be used to determine the presence of a corpus luteum (Fig. 3).

Repeated examination will reveal whether the corpus luteum is persistent. An appropriate therapy may now be applied, thus saving valuable time. Moreover the scanner also gives more detailed information about follicular activity.

Preovulatory follicles.

Manual rectal examination yields information about follicular changes during oestrus. The detection of these changes may be important in order to predict the day of ovulation as accurately as possible. Sound advice may now be given as to the optimal time for mating or insemination. This prediction is based upon the fact that approximately 12 h before ovulation the preovulatory follicles have a soft consistency. In approximately 10% of the cases this criterium does not hold (Ginther, 1986).

Echographic examination has also shown that, during the last days before ovulation, changes in the shape of the follicle may take place (Ginther and Pierson, 1984b; Ginther, 1986; Palmer and Driancourt, 1980): the shape changes gradually from round to conical (Fig.1). These changes in the preovulatory follicle may be seen in about 85% of the cases. A combination of both these criteria, consistance and shape, will result in a more relaible prediction of the moment of ovulation.

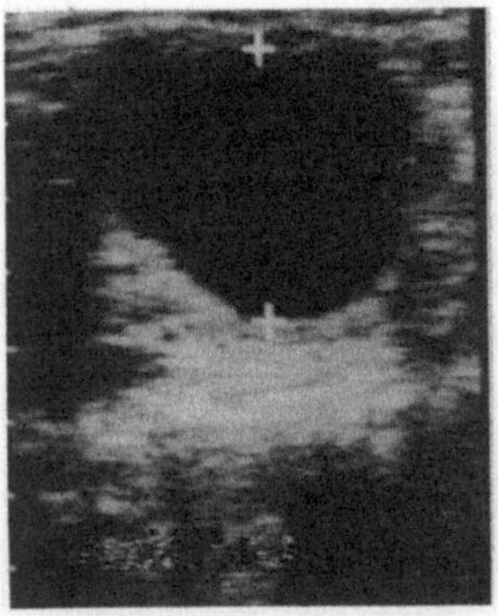

1. Pre-ovulatory follicle

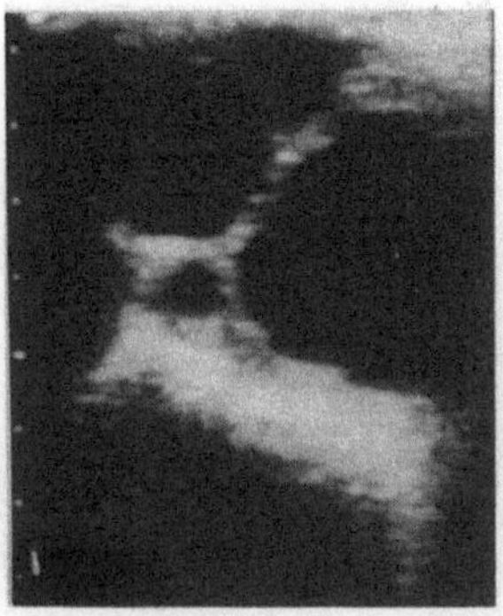

2. Double follicle

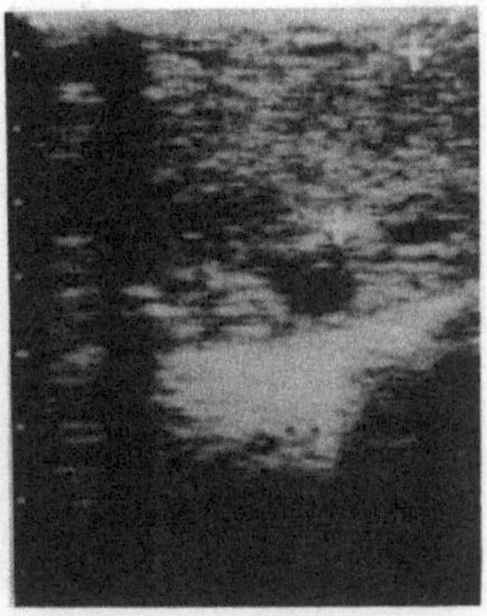

3. Corpus luteum

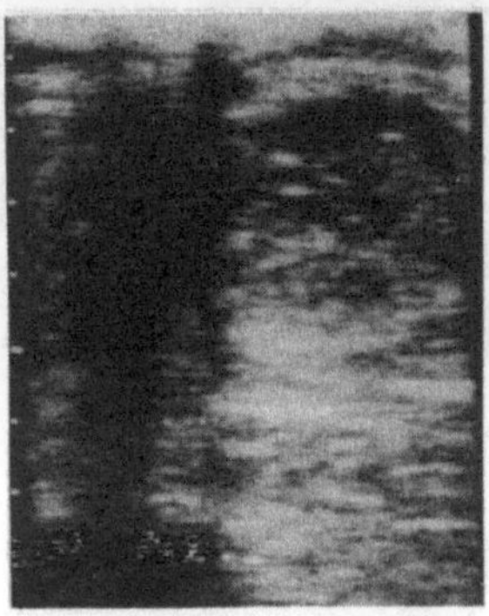

4. Luteinized follicle

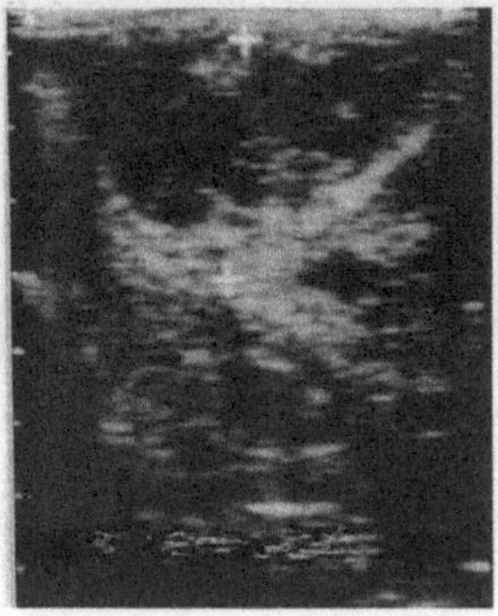

5. Corpus haemorrhagicum

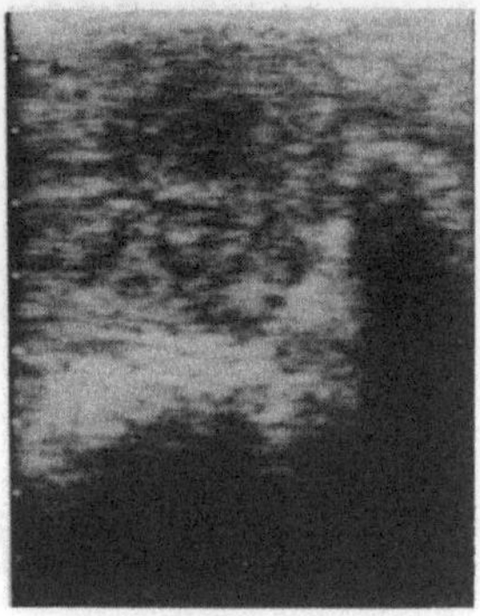

6. Uterus: under estrogen influence

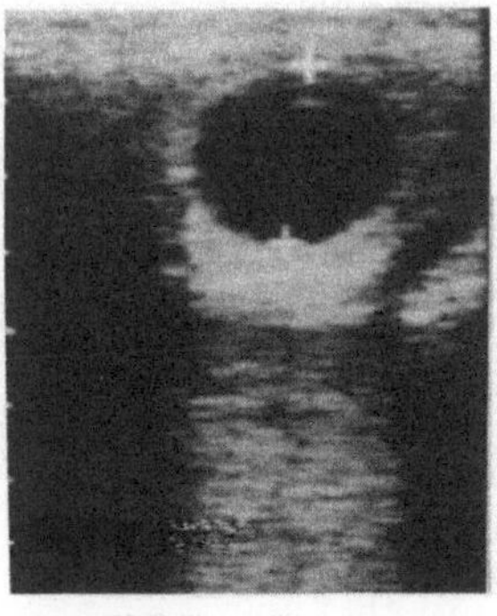

7. 18 days pregnancy

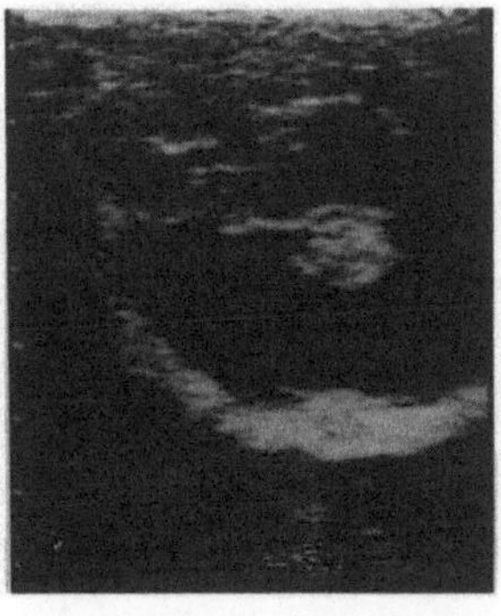

8. 40 days pregnancy

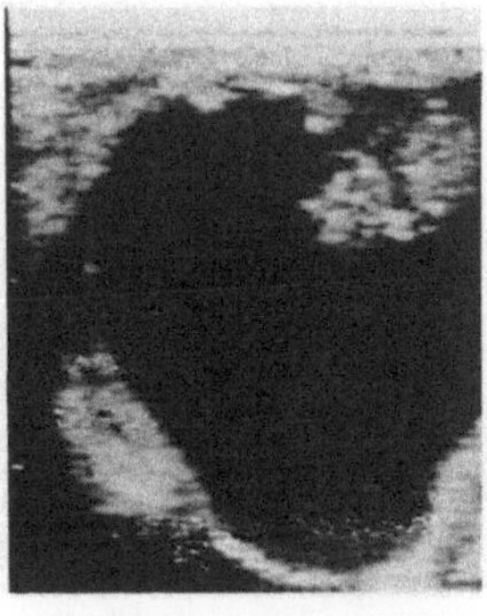

9. 6weeks twin pregnancy

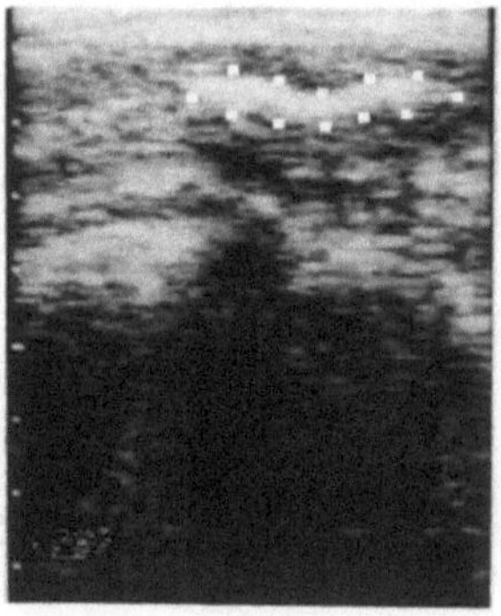

10. Uterus with "lost" swab

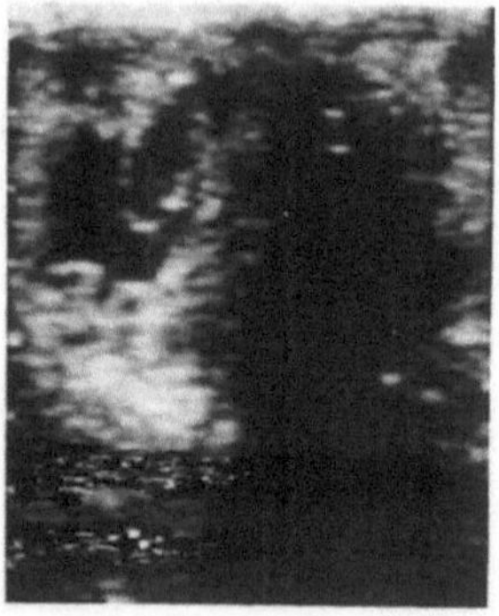

11. Endometrial cyst

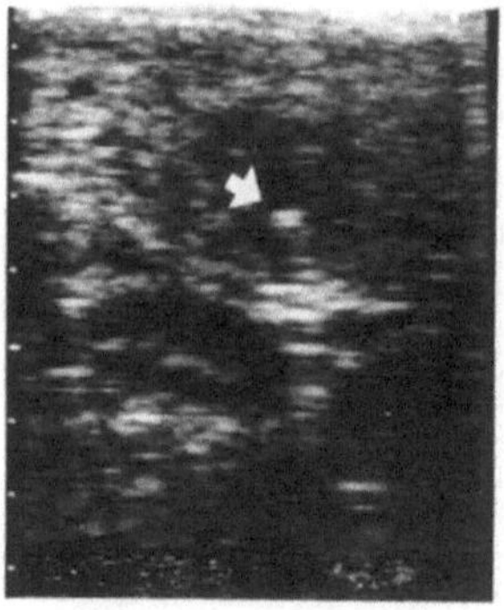

12. Uterus with small bone (maceration)

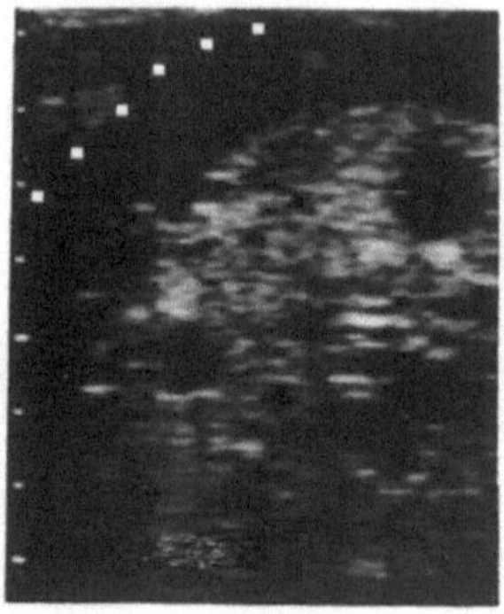

13. Ovarian tumor

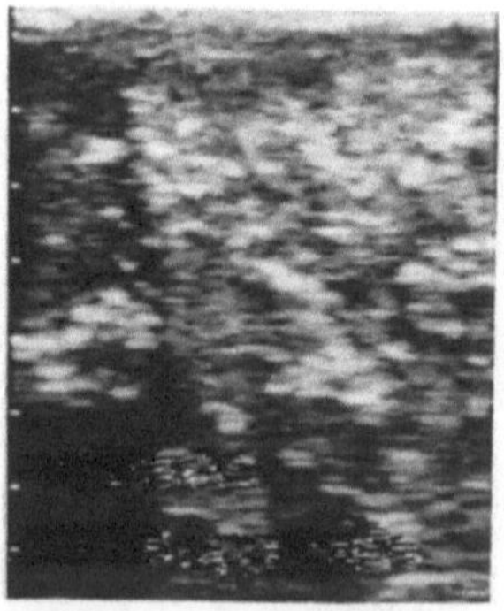

14. Ovarian absces

Multiple ovulations

During each period of oestrus more than one ovulation may occur. Follicles may ovulate concurrently, on the same day, or else on different days. Determining double ovulations may be important in connection with the possibility of twin pregnancies. It is often possible to palpate more than one preovulatory follicle, except when they are situated close to each other in the ovary (Ginther and Pierson, 1984b).

Using the scanner it is possible to determine the presence of a number of preovulatory follicles with greater certainty. Moreover, regular echographic examination, whereby the images are recorded, makes it possible to determine whether or not the presence of numerous preovulatory follicles will indeed lead to multiple ovulations, or whether the superfluous follicles have atrophied before ovulation (Fig. 2).

Corpus haemorrhagicum versus persisting follicle.

After ovulation during a normal oestrus, it is possible to determine the disappearance of the preovulatory follicle by rectal palpation. Sometimes, however, the symptoms of oestrus diminsh while the preovulatory follicle seems to persist. It is difficult to distinguish a persisting follicle from a firm corpus haemorrhagicum by rectal examination alone.

During echographic examination a corpus haemorrhagicum may be recognized by its homogenous grey, echogenic structure (Fig. 5). The content of a persisting follicle, on the other hand, is clear. In the case of luteinization a "cloudy" thickening of the wall is discernable (Fig. 4).

Pregnancy versus pseudo-pregnancy.

Both the absence of signs of oestrus and the manual determination of an increased tonus on the uterus may indicate pregnancy.

These same findings however may also be present in the case of early embyronic death followed by persisting corpus luteum. Afer a late mid-cycle ovulation too, the mare will not show signs of oestrus at the expected time while the tonus on the uterus will be increased.

The differentiation between pregnancy and these forms of pseudo-pregnancy can only be made with the help of echographic examination aimed at detecting the presence of absence of a foetal vesicle.

Ovarian tumours.

A great deal of information about abnormalities of the ovary, such as the presence of cysts, tumours, abcesses and abnormal peri-ovarian structures, may be obtained by manual rectal examination. However, sometimes for the therapy or surgical approach it is necessary to use ultrasonography in order to obtain additional information about these abnormalities (see also the contribution of Leidl and Kähn).

When a surgical approach is decided upon, it is important to know whether the tumour is solid or cystic so that the location of the laparotomy and length of the incision may be decided upon. In the case of a cystic tumour a large part of the contents may be aspirated so that the incision in the abdominal wall will be much smaller (Fig. 13 and 14).

Non-pregnant uterus.

Changes in the consistency of the uterus during the cycle and the possible presence of abnormal contents may be diagnosed by manual rectal examination.

However this method is subjective and interpretation is often quite difficult. The finer details of the uterine wall and the possible contents cannot always be determined in this way.

By scanning the uterus the consequences of oestrogen dominance can be visualized. During oestrus the echo-structure of the uterus is characterized by alternatively hypo- and hyperechogenic areas, representing oedema in the endometrial folds (Fig.6).

Whenever the results of oestrus detection and rectal examination do not agree, ultrasonographic examination may provide better insight.

Endometrial cysts.

If large enough, the fluid-filled endometrial cysts may be recognized by rectal examination (Kenny and Ganjam, 1975). These cysts, particulary when they are present in large numbers, may have consequences for the conception and pregnancy. They may protude into the luman or even be stalked; most are multilobular.

Endometrial cysts, even when small, are easier to locate using a scanner. Solitary cystic structures may easily be confused with the foetal vesicle (Fig 11).

Pregnant uterus.

Pregnancy may be diagnosed at an early stage by rectal examination (Bain, 1967; van Niekerk, 1965). On day 21 the tonus on the uterus is slightly increased and the location of the foetal vesicle in the uterus is as large as a walnut. At 5-6 weeks the diagnosis may be made with more certainty. The tonus is more pronounced and the location of the foetal vesicle is by now as large as an orange. The cervix is firm, long and thin.

Using a scanner a much more reliable pregnancy diagnosis may be made at a much earlier stage. After only 11 days post ovulation it may already be possible to detect the foetal vesicle.

In practice, pregnancy diagnosis is frequently performed after 21 days; usually because the owner wishes to have some certainty at this stage.

If it is known that a double ovulation has taken place, scanning at this time may be very useful (see Twins).

The signs of pregnancy may not be very pronounced during manual examination after 21 days and after 5 weeks, for example if there is too little tonus on the uterus and the cervix is too broad. Despite this, the mare may be pregnant. Ultrasonographic examination may provide an unequivocal answer (Fig. 7+8). At this stage it may even be possible to see the beating heart of the embryo.

Diagnosis of early embyronic death.

On the basis of rectal examination an incorrect diagnosis of embryonic death is frequently made. The absence of oestrus symptomes after three weeks and a negative pregnancy diagnosis on day 35 is usually interpreted as early embryonic death. Even if after examination on day 21 the pregnancy diagnosis seems to be positive because of an increased tonus on the uterus, but is negative on day 35, the diagnosis of early embryonic death cannot be made with certainty.
In both cases a mid-cycle ovulation may have taken place, which would explain the absence of oestrus and increased uterine tone.

A definite diagnosis can only be made by repeated echographic examination. The diagnosis is then based on either finding a normal foetal vesicle which during a subsequent examination shows signs of degeneration or has completely disappeared, or on the disappearance of the previously detected foetal heart action.

Pregnancy duration.

Transcutaneous scanning may be used to determine the size of the various foetal organs. As such it is a useful aid for estimating the age of the foetus in a later stage of pregnancy (Rossdale, 1988). Transcutaneous scanning is also applied to small ponies when transrectal scanning is impossible.

Twin pregnancies.

Multiple ovulations need not be alarming. Only a quarter of the double ovulations will result in twin pregnancies. Many of these twins will be reduced to single pregnancies because of the death of one conceptus in an early stage of the pregnancy. However an early diagnosis of twin pregnancy is still important. Foetal vesicles may be detected by rectal palpation from day 20, particulary when the localisation is bilateral. The sooner the diagnosis is made the better, at any rate before the development of the endometrial cups around day 35.

Ultrasonography is an excellent method to diagnose twin pregnancy. Bilaterally localized twins are easy to find. However, even with the help of a scanner, twins fixed unilaterally may be difficult to detect in an early stage of pregnancy (Fig. 9).

Repeated echographic examination may reveal the death of one of the conceptuses (Ginther, 1986; Simpson et al., 1982).

A twin prevention program must be started early. When a double ovulation has taken place, the search for a possible twin pregnancy should be started between days 12 and 14 after ovulation. Should two foetal vesicles be present, one of them should be seperated by pushing it, using finger and thumb, to the top of the uterine horn and then crushing it. Repeated examination of the remaining vesicle is essential. Even after the fixation, one of the vesicles of a bilaterally situated twin may be manually crushed. Reducing twins fixed unilaterally is more difficult at this stage. In this case inducing an abortion would be a better solution.

ULTRASONOGRAPHY DOES NOT YIELD MORE INFORMATION THAN RECTAL MANUAL EXAMINATION.

Ovarian cysts

In most cases ovarian cysts are easily diagnosed by rectal examination. The scanner is useful in cases where more information concerning the thickness of the cystic wall is desired.

Endometritis

Even though the presence of small amounts of fluid in the uterus, suggestive of inflamation, may be visualized with the scanner, endometritis may easily be diagnosed by a combination of manual rectal and bacteriological examination.

Pyometra

Rectal and vaginoscopic examination provide sufficient information to diagnose pyometra (Hughes et al., 1979). Echographic examination may provide the answer only when doubts arise between distinguishing early gestation from a developing pyometra.

Maceration

Although maceration of the equine foetus does not frequently occur, the diagnosis by rectal palpation of the uterus is not difficult to make.

Crepitation of the uterine contents between finger and thumb is clear; it may even be possible to palpate small bones through the uterine wall.

It may reassuring to use echography to check that the uterine contents has completely disappeared after treatment (Fig. 12).

CONCLUSION

An earlier and more reliable diagnosis of reproductive disorders of the mare allows for an early application of direct therapies. Where manual rectal examination is often sufficient, ultrasonography offers a wider range of possibilities. In practice, routine examination will generally be performed manually. When doubts arise, or more information is required, ultrasound scanning is indispensable.

Frequent use of the technique will improve the interpretation of the monitored images and will serve to enlarge the diagnostic knowledge. Ultrasonography has become an integral part of the reproductive management of mares.

REFERENCES

Bain, A.M. (1967) "The manual diagnosis of pregnancy in the thoroughbred mare", New Zealand Vet. J. 15, 227-230.

Ginther, O.J. and Pierson, R.A. (1984a) "Ultrasonic evaluation of the reproductive tract of the mare: Ovaries", Equine Vet. Sc. 4, 11-19.

Ginther, O.J. and Pierson, R.A. (1984b) "Ultrasonic anatomy of equine ovaries", Theriogenology 21, 471-483.

Ginther, O.J. (1986) "Ultrasonic imaging and reproductive events in the mare", Equiservices, Cross Plains, U.S.A.

Hughes, J.P., Stabenfeldt, G.H., Kindahk, H., Kennedy, P.C., Edqvist, L.E., Neely, D.P. and Schalm, O.L.W. (1979) "Pyometra in the mare", J. Reprod. Fertil. Suppl. 27, 321-329.

Kenney, R.M. and Ganjam, V.K. (1975) "Selected pathological changes in the mare uterus and ovary", J. Reprod. Fertil. Suppl. 23, 335-339.

Niekerk, C.H. van (1965) "The early diagnosis of pregnancy, the development of the foetal membranes and nidation in the mare", J.S. Afr. Vet. Med. Ass. 36, 483-488.

Palmer, E. and Driancourt, M.A. (1980) "Use of ultrasonic echography in equine gynaecology", Theriogenology 13, 203-216.

Rossdale, P.D., et al. (1987) "Transabdominal ultrasound scanning of the equine fetus from 100 days gestation to full term", Proc. Ultraschall beim Pferd- Intensivmedizin beim Fohlen, Bern.

Simpson, D.J., Greenwood, R.E.S., Ricketts, S.W, Rossdale, P.D., Sanderson, M. and Allen, W.R. (1982) "Use of ultrasound echography for early diagnosis of single and twin pregnancy in the mare", J. Reprod. Fert. Suppl. 32, 431-439.

ULTRASONIC CHARACTERISTICS OF PATHOLOGICAL CONDITIONS OF THE EQUINE UTERUS AND OVARIES.

LEIDL, W. AND W. KÄHN
Gynaecological and Ambulatoric Animal Clinic, University of Munich
Königinstraße 12
D-8000 München 22
Germany

Introduction

During the initial years of its use in veterinary medicine, ultrasound technology was applied primarily for the assessment of normal morphological features of the reproductive tract (Palmer and Driancourt 1980, Simpson et al. 1982, Valon et al. 1982, Ginther and Pierson 1984 a). Serving as a reliable instrument for pregnancy diagnosis and the examination of ovarian follicles and corpora lutea, its use spread rapidly. Pathological phenomena were usually incidental findings and were considered under the aspect of differentiating them from physiological characteristics (Ginther and Pierson 1984 b, Leidl and Kähn 1984).

Since then, increasing attention has been directed towards its use in the diagnosis of pathological conditions of the uterus and ovaries of the mare. In the following section echogenic properties of the images displayed by various pathological structures of the equine reproductive tract are described.

Pathological findings of the uterus

ENDOMETRIAL CYSTS

Endometrial cysts have been found mainly in mares over 10 years of age (Kenney and Ganjam 1975, Kaspar et al. 1987). These cystic structures are either lymphatic or glandular in origin. Ultrasonic images of these structures occasionally resemble the sonographic pattern of embryonic vesicles during early gestation (Fig. 1).

Endometrial cysts occur solitary or multiple (Fig. 2 and 4) and are located anywhere within the uterine body or horns. Sonographically they present the image of spherical to ovoid shaped, fluid-filled vesicular bodies. Cysts consist of a single cavity or, quite commonly, a number of fluid-filled compartments (Fig. 3). The external wall of endometrial cysts and in the case of multichambered cysts, the dividing walls, display the same echogenity as the uterine wall. When the ultrasound beam meets the walls at right angles, a highly echogenic image appears on the screen. The cavity, containing non-echogenic watery fluid shows up near-black on the screen. Cysts vary greatly in size from only a few millimeters to several centimeters.

Solitary endometrial cysts with diameters of 10-30 mm may possibly be confused with an embryonic vesicle around day 10 to 25 of gestation (Leidl et al. 1987). Identification becomes more easy as of week 4 of embryonic development. From this stage on, conclusive evidence for an intact pregnancy can usually be obtained from the depiction of the embryo and embryonic heart activity. Even at this stage, however, diagnosis may be difficult, especially in the case of twin pregnancies with the concurrent presence of endometrial cysts (Fig. 4).

Sound knowledge of the sonographic anatomy of the intact conceptus is required to dif-

M. M. Taverne and A. H. Willemse (eds.), Diagnostic Ultrasound and Animal Reproduction, 21–35.

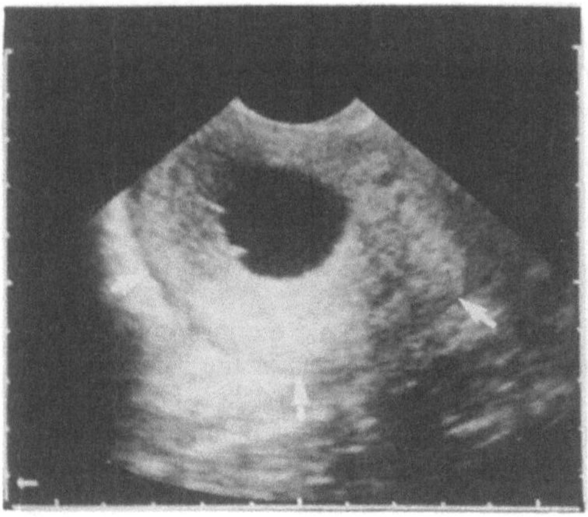

Fig. 1: Spherical singular endometrial cyst within the uterus (arrows) of a non-pregnant mare. The sonographic image resembles on early pregnancy around day 15.

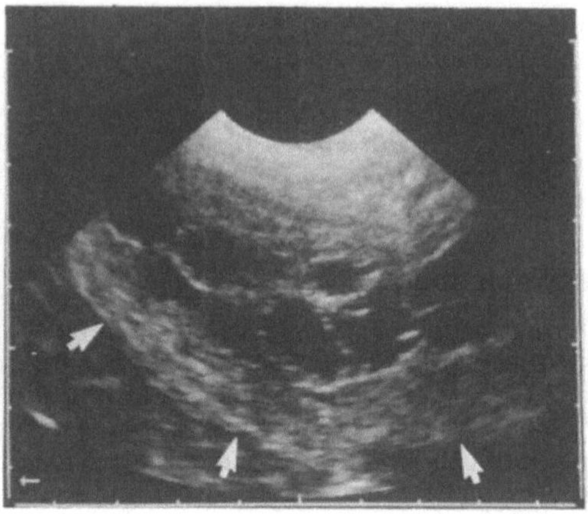

Fig. 2: Multiple endometrial cysts in the uterus (arrows) of a mare. The mare conceived a few weeks later and carried a normal foal full term.

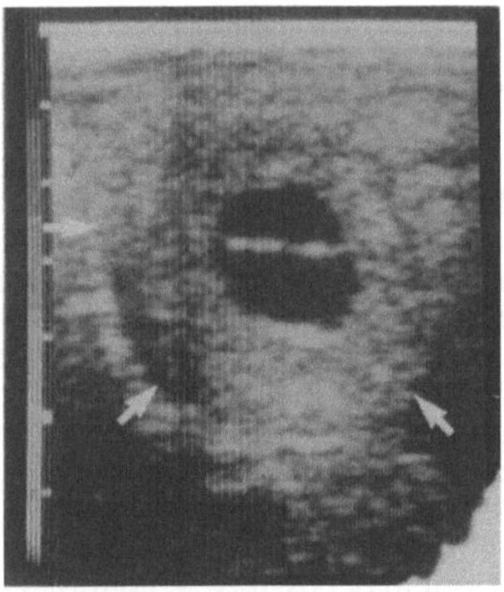

Fig. 3: Uterus (arrows) with a compartmentalized endometrial cyst which must be carefully differentiated from a pregnancy in the ascending phase. The wall between the compartments gives a similar appearance to the separating membrane between yolk sac and allantoic sac.

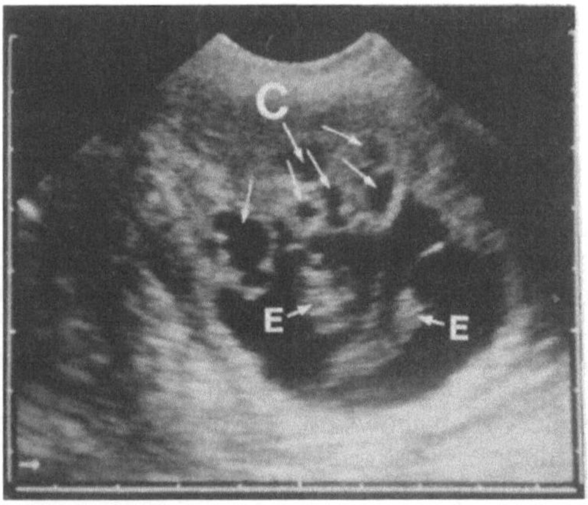

Fig. 4: Several endometrial cysts (C) present simultaneously with twin pregnancy. Both embryos (E) can be seen.

ferentiate it from cysts. At this stage, the embryonic vesicle is usually spherical in shape and normally located in the middle of the uterine lumen. The diameter of the conceptus should conform with developmental stage expected from the date of service. Two of the most reliable indices for an intact pregnancy are the mobility of the embryonic vesicle within the uterine lumen and embryonic heart activity. The mobility of the conceptus is particularly marked until day 15-16, and can often be observed during an ultrasound scan of a few minutes (Leith and Ginther 1985).

Indications of endometrial cysts are the presence of usually multiple, ovoid, irregular or compartimentalized fluid-filled structures in the uterus. Cysts are located excentrally in the uterine lumen or within the uterine wall. As mentioned above, the diameter of the vesicular bodies under investigation is important. In the case of a pregnancy, this should comply with the history of mating. In questionable cases, repeated scans a few days apart should reveal the absence of growth in the case of endometrial cysts.

In order to minimize later confusion, it is advisable to register the presence, number, location and size of endometrial cysts in mares prior to breeding. This is best done in conjunction with routine prebreeding-examinations before mating season starts. Should an additional vesicular body appear during the first three weeks after service, with the expected diameter of a conceptus of this age, the mare may well have become pregnant. In contrast to the embryonic vesicle, endometrial cysts remain virtually unchanged sonographically for some weeks.

EMBRYONIC DEATH

Various signs revealed by ultrasound scanning have been found to be indicative of possibly imminent embryonic death. Generally, all findings not anticipated under physiological conditions should make the examiner alert. There are, however, a number of defined changes which nearly always take place prior to the loss of an embryo.

The line of demarcation between the embryonic vesicle and the neighbouring endometrium is usually smooth. Waviness of this line gives one possible indication of pathological changes (Fig. 5). This is caused by the loss of internal fluid pressure of the vesicle due to cessation of production and beginning resorption of placental fluids. As dehydration progresses, the boundary between the vesicle and the endometrium becomes lobulated. The endometrial folds, which are normally distended by the fluid-filled vesicle, now begin projecting into the embryonic vesicle (Squires et al. 1988).

Changes in the shape of the embryonic vesicle are an important sign for complications in pregnancy. In the case of an intact pregnancy, the conceptus displays a perfectly spherical shape until day 15 or 16 of gestation. Any deviation from this may signalize imminent embryonic death (Fig. 7). After day 16, however, changes in shape occur physiologically and should be interpreted with care.

The echogenity of the placental fluids is another important criterion in judging the soundness of a pregnancy. With beginning resorption, echogenic particles appear in the usually non-echogenic vesicular lumen (Fig. 6 and 7). Increased echogenity is caused by an accumulation of cellular components and disintegration of the foetal membranes.

Evidence of embryonic heart activity remains a most reliable parameter for the viability of the embryo. Shortly prior to death, bradycardia with a heart rate of 100 beats per minute or less can sometimes be determined by ultrasound. For comparison, an intact embryo should have a heart rate of at least 150 beats per minute.

The size of an embryonic vesicle is usually smaller than expected in the case of disturbed pregnancy (Ginther et al. 1985). If resorption occurs in the first three weeks of pregnancy, it is often very quick. Only a few days usually pass between the first sonographic indications of irregularities and the disappearance of the conceptus. Delayed resorp-

Fig. 5: Disturbed pregnancy on day 27. The embryo showed heart activity but signs for impending embryonic death were detectable. These are the endometrial folds (arrows) bulging into the embryonic vesicle, the reduced amount of placental fluids and the small-for-age size of the conceptus.

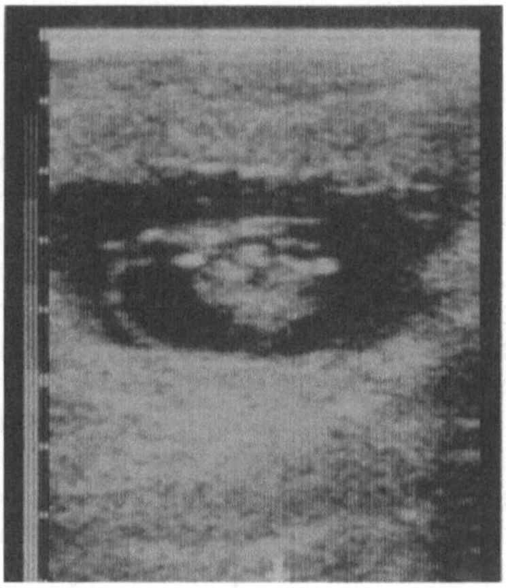

Fig. 6: The disturbed pregnancy depicted in Fig. 5 on day 33. The embryo had continued to grow and still showed heart activity. Reflections can be seen within the embryonic fluid. The embryo was dead 3 days later.

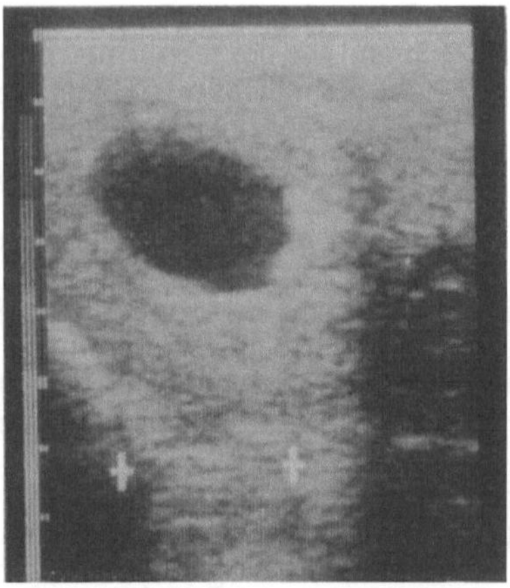

Fig. 7: Disturbed pregnancy on day 16. The embryonic vesicle has lost its spherical shape and weak echoes have appeared in the placental fluids. Five days later the embryonic vesicle has been almost completely resorbed.

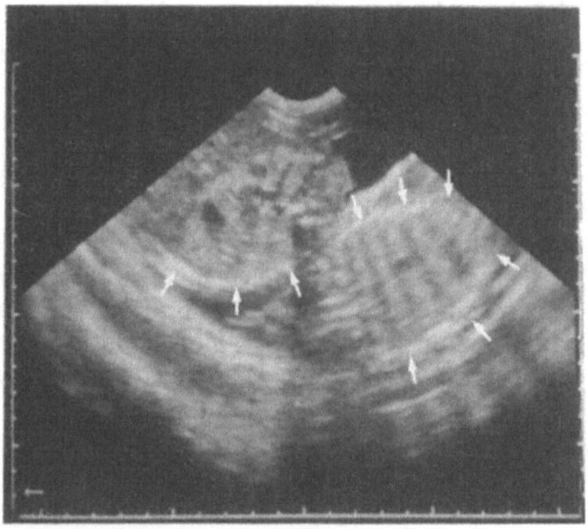

Fig. 8: Pathological changes in a twin pregnancy on day 129. Both foetuses had a heart rate of less than 80 beats per minute. The mare aborted the following day. Adjacent rib cages (arrows) of both foetuses can be seen.

tions may take place as of the third week (Fig. 5 and 6). An early sign of loss of conceptus is the small-for-date size of the embryonic vesicle. The embryo itself may continue to grow for a few days, sometimes even weeks (Darenius et al. 1988). These embryos do not, however, reach normal growth rate and they ultimately die.

FOETAL DEATH

Various signs for complications which result in the loss of the foetus in advanced stages of pregnancy can be detected by ultrasound scanning. The size of the allantoic sac, having reached a diameter exceeding the field of view of the ultrasound beam, can no longer serve as an criterion for these examinations. Now the foetus itself becomes relevant. Regrettably, knowledge of the sonographic anatomy of the equine foetus is still quite sparse. However, recent reports do provide valuable information on complications in advanced pregnancies (Pipers and Adams-Brendemuehl 1984, Kähn and Leidl 1987 b).

It was shown that malconditioned foetuses had a definitely lower heart rate than sound foetuses. Thus, examination of a twin pregnancy revealed a heart rate below 80 beats per minute for the two foetuses on day 129 of gestation (Fig. 8). Both foetuses were aborted the following day. For reference, the normal frequency is 100-180 beats per minute at this stage of pregnancy (Colles and Parkes 1978, Matsui et al. 1985).

In another case, the still viable foetus showed clearly dilated blood vessels in the abdominal region a few days prior to abortion. The echogenic impression was that of very broad non-echogenic bands which were particularly prominent in the liver. This pattern was interpreted as a sign of congestive circulatory problems.

Often, the foetus is not immediately aborted after death. Repeated ultrasound scans reveal a steady reduction in placental fluids. Furthermore, the anatomy of the foetal organs changes. Structures which are very distinct in the living foetus become blurred or obscure after death (Fig. 9). This is especially true for soft tissue organs which undergo rapid postmortem changes (Staudach 1986). As a result of intravascular blood coagulation, blood vessels which are normally echo-poor, become difficult to discern from the surrounding tissue. The somatic order usually found in these organs is lost and they appear uniform on ultrasound investigation. Ossified structures, such as bone, being hyperreflective, do remain detectable for some time. Thus, foetal echoes within the uterus have been scanned successfully several weeks after death, when all the placental fluids had been resorbed (Ginther et al. 1985).

POSTPARTUM UTERUS

Nearly all mares show ultrasonically detectable collections of fluids within the uterus during the first 5 days after parturition. Occasionally, an ultrasonic scan a few days postpartum may give the impression that the uterus contains no such fluids (Fig. 10). Frequently, however, repeated examinations of these mares 1 or 2 days later reveal the presence of lochia (McKinnon et al. 1988). The percentage of mares still having uterine fluids decreases as of day 7 postpartum. Intrauterine fluids could be demonstrated in 25% of the mares under investigation at the time of postpartum oestrus. The echogenity of the lochia is initially relatively high and steadily decreases during the first 2 weeks postpartum. Directly following parturition it contains numerous flaky echoes displayed by discarded tissue matter and inflammatory products. Viscous, pussy lochia are highly echogenic.

An important finding was that among mares studied after foaling, only about one third of those which had shown intrauterine collections of fluid at the time of service, became pregnant (McKinnon et al. 1988). On the other hand, 80% of the mares with good uterine involution became pregnant. The practical consequence for the reproductive management

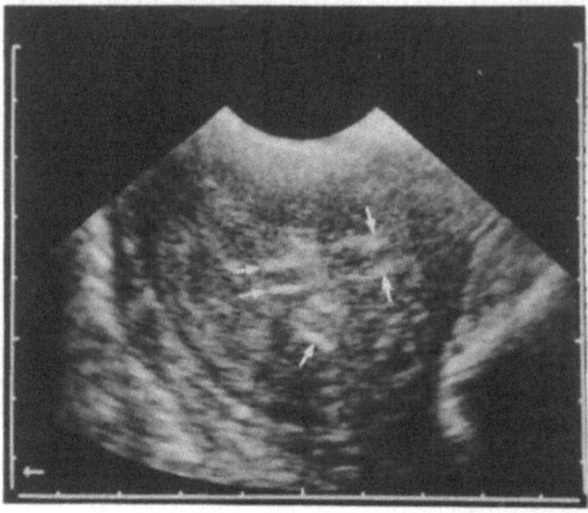

Fig. 9: Ultrasound image of an uterus 4 days after the death of a foetus around day 70 of gestation. The placental fluids have almost completely disappeared. Highly echogenic foetal parts (arrows) were sonographically detectable for further two weeks.

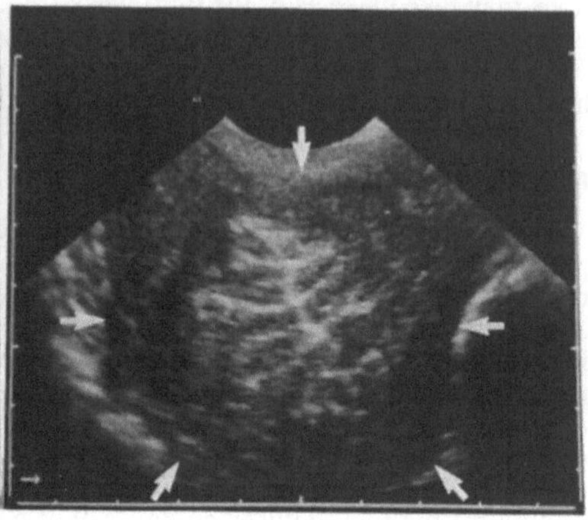

Fig. 10: Normal involution of the uterus (arrows) in a mare 3 days after parturition. The highly echogenic branching pattern in the center is caused by the surface of apposing endometrial folds.

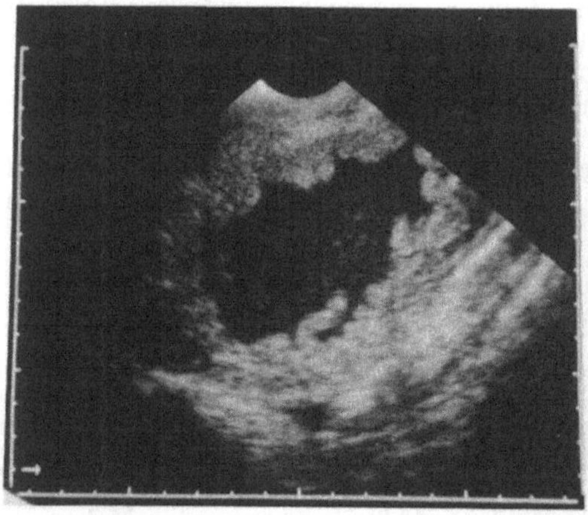

Fig. 11: Uterus containing large amounts of fluids 4 days post partum in a mare with puerperal disorders.

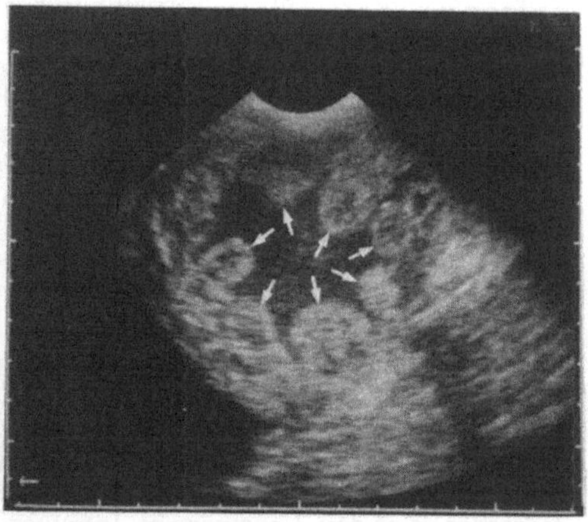

Fig. 12: Collections of scattered echogenic exudate in the uterine lumen in a case of chronical endometritis. Note the meandering endometrial folds (arrows).

of mares based on these results, is that only mares with no ultrasonically detectable lochia should be served during postpartum oestrus. Lochia were found within the uterus in all ultrasonically examined cases of postpartum complications involving uterine inflammation (Fig. 11).

ENDOMETRITIS

Sonographic images of chronic endometritis are marked by the presence of abnormal collections of fluid in the uterine lumen. The amount of fluid varies considerably among individual mares, as well as between consecutive examinations of the same mare a few days apart. An important factor of consideration is the stage of the oestrus cycle. Collections of fluids can occur physiologically during normal oestrus (Adams et al. 1987). Therefore, only if such fluids are encountered during dioestrus, it is justified to consider a pathological origin (Fig. 12). The fluids may be restricted to a few locations within the uterus, or they may be demonstrated along the entire uterine lumen.

Ultrasonically, these collections of fluid are characterized by a meandering boundary (Leidl and Kähn 1984). Endometrial folds can be observed protruding into the lumen on a cross-sectional sonographic plane (Fig. 12 and 14). Frequently all 6-8 endometrial folds which physiologically occur can be identified. The pathological fluids found in the uterus fill out the spaces between the folds. These are normally not distinguishable during dioestrus as they lie adjacent to one another with no space in between the folds.

The meandering boundary of the fluid collections is characteristic for endometritis distinguishing this pathological disorder from other fluid-filled structures in the uterus. Internal fluid pressure of an embryonic vesicle or endometrial cysts causes these vesicular bodies to become well-rounded, bulging into the uterine wall and straightening the mucosal folds. In cases of pregnancies two possible causes for a meandering border between the fluid-filled structure and the uterine wall are resorption of amnio-allantoic fluids or foetal death. There are, however, exceptional cases, when intact embryonic vesicles may also show a wavy border. In all cited cases, the mares were aged and had foaled several times before.

The typical echogenic properties of the fluids found in cases of endometritis may be useful for differential diagnosis. Endometrial secretions contain numerous particles of varying echogenity. The echogenic properties of the image can range from scattered bright flakes to high echogenity, brighter than that of the uterine wall.

However, it is important to realize that the echogenity of the amniotic fluid also increases during the second and final third of pregnancy. Due to an increase in cellular content, initially flaky, later snow-storm effects are produced. In the final third of pregnancy, this also holds true for the allantoic fluid.

From the above, it becomes clear that the differentiation of fluids found in the uterus due to endometritis from early stages of gestation and endometrial cysts requires careful consideration of various aspects. Several criteria of importance are echo-pattern, shape and intrauterine location of the fluid collections in question. It has been found that the optimal time to diagnose endometritis sonographically is late dioestrus (Adams et al. 1987). The pathological fluid secretions are most evident at this stage and will not be confused with secretions which may occur during oestrus. Furthermore the period around day 14 to 16 is the best time for early pregnancy diagnosis with the additional advantage of a promising intervention in the case of a twin-pregnancy.

Pyometra can be considered a special form of endometritis. The ultrasonic impression is characterized by an extreme dilation of the uterus (Fig. 13). This distends the uterine wall causing the endometrial folds to disappear and the boundary between uterine content and uterine wall becomes smooth. Pyometra is usually recognized by intense echo-patterns

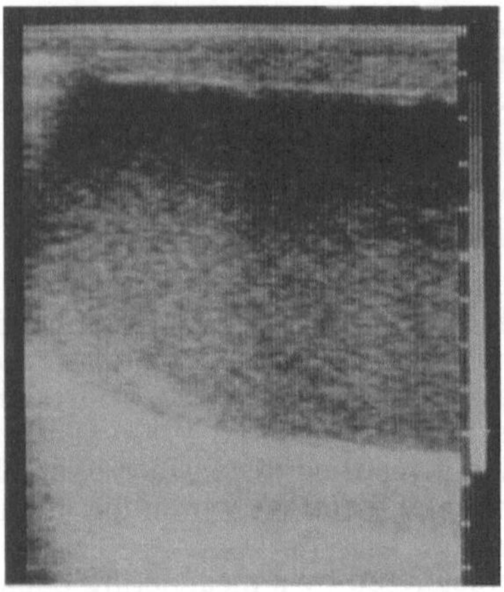

Fig. 13: Large intrauterine collections of fluid attributable to a pyometra. The purulent material shows intensifying echoes towards the bottom of the uterus due to the increasing amount of cellular particles.

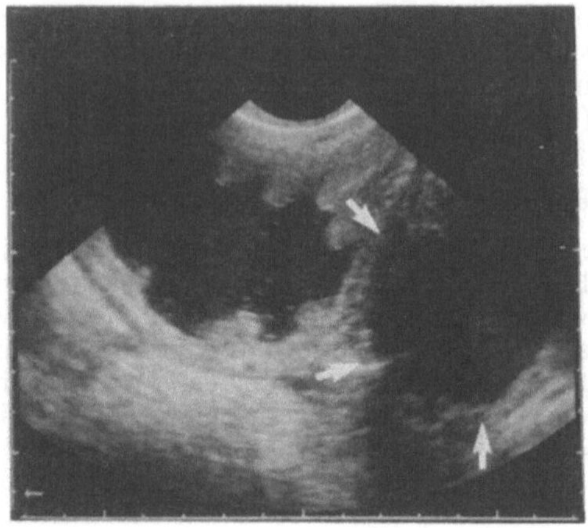

Fig. 14: Severe endometritis in a mare. The fluid-filled uterus with protruding endometrial folds lies cranial to the urinary bladder (arrows).

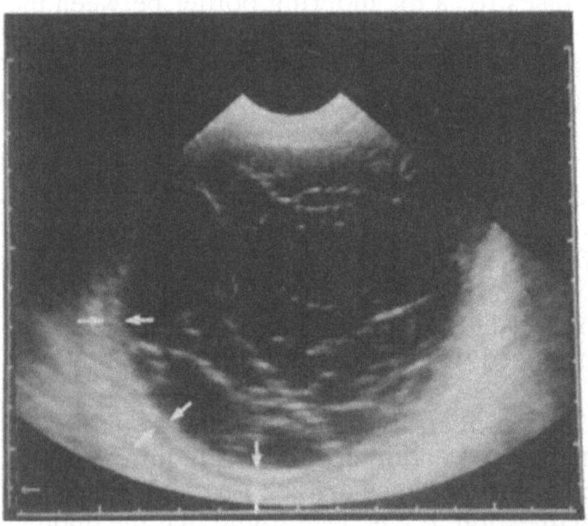

Fig. 15: Anovulatory follicle with netlike internal echoes in a mare. The follicular wall (arrows) can be seen. The mares plasma progesterone value was 5.1 ng/ml.

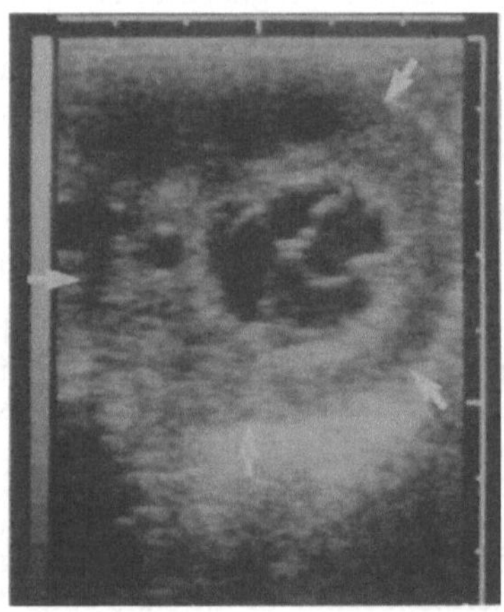

Fig. 16: The anovulatory follicle depicted in Fig. 15 examined 7 days later shows a sonographic image resembling a corpus haemorrhagicum. The nonechogenic central zone is surrounded by an echogenic wall (arrows). The progesterone value of the mare was 7.6 ng/ml plasma.

which become increasingly dense towards the ventral aspect of the uterus. Echogenity is caused by cellular particles suspended in the secretions which become more concentrated ventrally in the viscous purulent mass.

The sonographic anatomy of the equine urinary bladder can be confused with that of a pyometra (Fig. 14). Due to the particular consistency of equine urine, the lumen of the bladder shows intense echoes, similar to pyometra. The shape of the bladder can also resemble the cranially dilated uterus. In order to minimize diagnostic error, a pyometra should be diagnosed only if both, the uterus and the bladder, can sonographically be demonstrated.

If the sonographic differentiation of a pyometra from an advanced pregnancy presents problems, the floating amniotic membrane, parts of the foetus or the umbilical cord must be looked for.

Pathology of the ovaries

ANOVULATORY, LUTEINIZED FOLLICLES

Although a follicle may grow during oestrus to become the dominating follicle, ovulation sometimes fails to take place. Such anovulatory follicles may have the size of normal preovulatory follicles (Will et al. 1988). Frequently, however, they grow further to achieve diameters of up to 6-10 cm, in rare cases even more. During oestrus, the sonographic image of anovulatory follicles does not differ from that of normal follicles. Some of these anovulatory follicles appear to develop into haemorrhagic follicles, others show tendencies to become luteinized (Squires et al. 1988).

Just after the stage when ovulation should have taken place, early changes in the form of increased echogenity within these anovulatory follicles may occur (Fig. 15). The normally echo-poor, fluid content of the follicles becomes dispersed with flaky echoes or network-like echo patterns. These reflections may be displayed by blood components as appear in haemorrhagic follicles or they represent first signs of luteinization as in luteinized follicles. Once anovulatory follicles show these phenomena, in most cases oestrus is over and the mare refuses the stallion.

Soon after the first appearance of scattered internal echoes, luteinizing follicles show an increase in thickness of the originally fine echogenic follicular wall. Further development then resembles changes which occur in haemorrhagic corpora lutea during dioestrus (Fig. 16). These show an increase in diameter of the originally narrow band of luteal tissue surrounding the haemorrhagic core, which therefore becomes smaller (Pierson and Ginther 1985, Kähn and Leidl 1987 a). Examinations have demonstrated that luteinization of anovulatory follicles occasionally leads to corpus luteum-like structures. Increased progesterone values of the affected mares seem to correspond to these morphological changes. Thus, a few cases have been found, where high levels of progesterone were determined in mares with anovulatory follicles showing increased echogenity. The following oestrus began on the expected date.

Developmental changes similar to the luteinization of anovulatory follicles during oestrus, could also be observed during early pregnancy (Fig. 17). Around day 40 to 50 of pregnancy, a time when accessory corpora lutea are expected to develop, the ovaries of pregnant mares showed large follicles with netlike internal echoes. Sonographic follow-up examinations during the following weeks revealed a steady increase in echogenity (Fig. 18). The cavity of these follicles gradually became occupied by tissue with the same echogenic properties as corpora lutea. These structures could therefore have been corpora lutea accessoria.

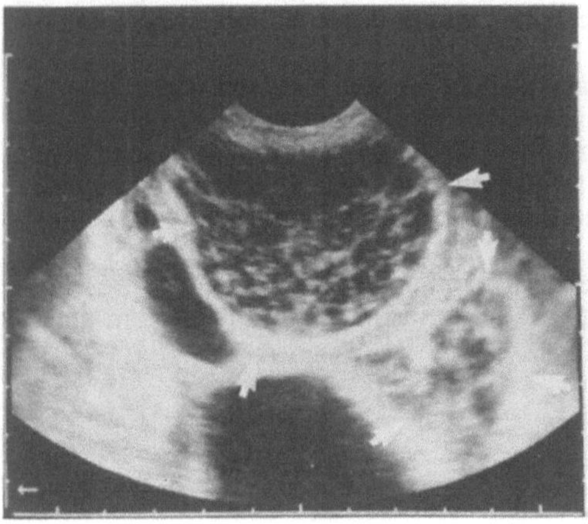

Fig. 17: Two corpora haemorrhagica (arrows) accessoria in a mare on day 60 of pregnancy.

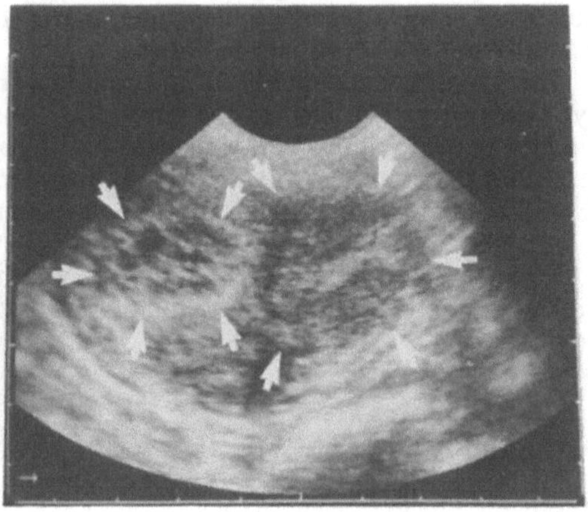

Fig. 18: The two accessory corpora lutea of pregnancy depicted in Fig. 17 had become much more compact and appeared luteinized (arrows).

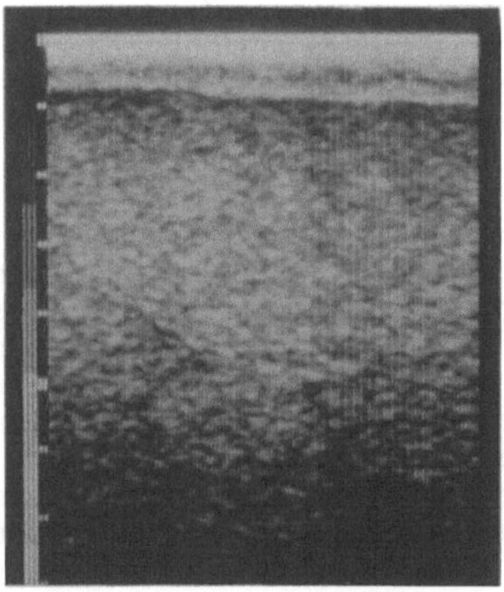

Fig. 19: Sonographic section of a large ovarian haematoma. The uniform granular echogenity is typical for the sero-haematic content.

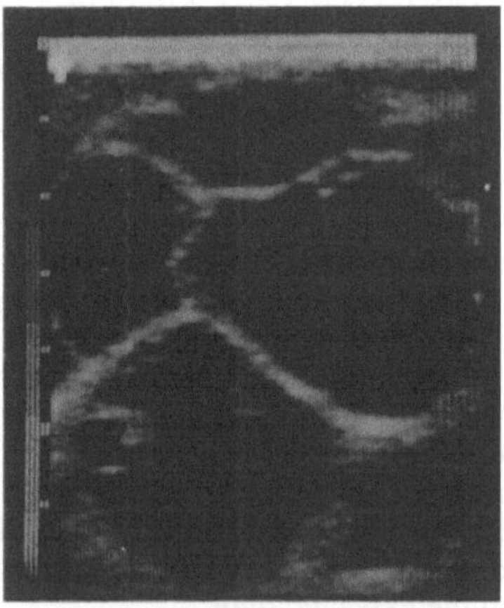

Fig. 20: Sonographic section of the ovary of a mare with large cystic ovarian degeneration. The thin cystic walls are highly echogenic. Functional ovarian tissue had become atrophic and was no longer visible.

FOLLICULAR AND OVARIAN HAEMATOMAS

The results of studies performed so far suggest that the majority of anovulatory follicles develop into structures sonographically resembling corpora lutea. The remainder of anovulatory follicles is subjected to diapedesis bleeding and the follicular cavity becomes filled with blood. No luteinization as described above occurs. This type of anovulatory follicle is designated haemorrhagic follicle or follicular haematoma (Squires et al. 1988). If the remaining ovarian tissue becomes atrophic and the haematoma encompasses almost the entire ovary, it is referred to as an ovarian haematoma.

Follicular haematomas do not show a sonographically detectable increase in breadth of the peripheral echogenic zone, an expression of progressing luteinization (Fig. 19). In contrast, the wall of these anovulatory follicles remains a thin echo-intensive band during several days after oestrus. The follicular cavities do not show network-like echogenic patterns. Rather, the relatively non-echogenic interior zone of these follicles later becomes interspersed with snow-flake-like reflections. These progressively become more dense giving the "snow-storm" effect. A few bright echogenic lines could be detected in rare cases. These are interpreted as fibrinous strands of the clotted blood and the beginning organization of the haematoma.

Anovulatory follicles which develop into follicular haematomas often have the diameter of a mature ovulatory follicles. Sometimes, however, these follicles increase considerably in size after oestrus. In one mare, for example, the vesicular structure on the ovary under investigation developed into an ovarian haematoma with a diameter of more than 20 cm during the four weeks following oestrus (Fig. 19). The excised ovary weighed 3.8 kg and consisted almost entirely of a single haematoma cavity filled with sero-haematic fluid.

The ultrasonic image of this ovarian haematoma was that of an extremely large vesicular structure whose cavity was various shades of grey with dispersed small bright echoes. No signs of internal organization in the form of definite echo-intense lines, were detectable.

OVARIAN CYSTS

Ovarian cysts - that means sonographically follicle like structures which do not change during a longer period of time - are not common in mare but they do occur. In one case, a 13-year-old mare had shown irregularities of the oestrus cycle over a period of 2 years and had been served unsuccessfully several times. Detailed clinical examinations including repeated rectal palpation, hormonal and sonographic investigations gave evidence for the presence of ovarian cysts. As these structures failed to react to hormonal treatment, the mare was unilaterally ovariectomized. Pathological examination revealed large cystic degeneration of the ovary, with atrophy of the functional tissue. There were no indications of a blastoma.

Sonographic scans of this large cystic degeneration of the ovary were characterized by numerous, large vesicular structures separated by thin, echogenic walls (Fig. 20). The cysts appeared polygonal in cross-section and their content was almost echo-free. No echogenic areas of tissue could be detected between the follicles. In contrast, ultrasonography of intact ovaries with follicles shows areas of echo-rich, organized, functional ovarian tissue between the follicles.

OVARIAN TUMOURS

Ovarian tumours of the mare may present a variety of sonographic impressions. Among the granulosa-cell tumours, which are reported to be the most common ovarian neoplasm in the mare, multichambered cystic blastomas are found most frequently (White and Allen 1985, Kähn and Leidl 1987 a). An ultrasound scan will reveal the presence of far more

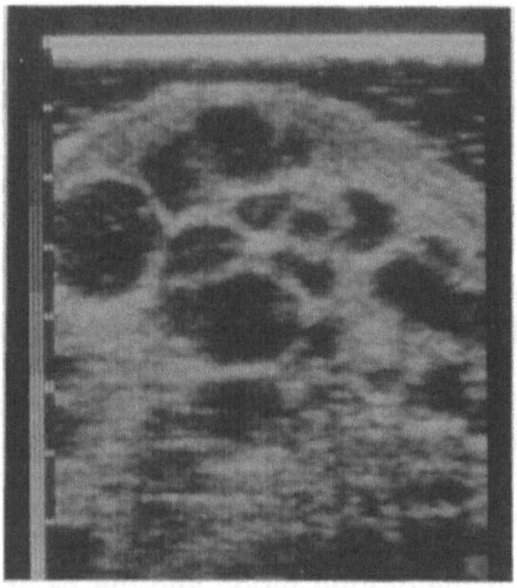

Fig. 21: Sonographic section of an equine ovary with a granulosa-cell tumour. The image displays a part of the typical multicystic structure of the tumour.

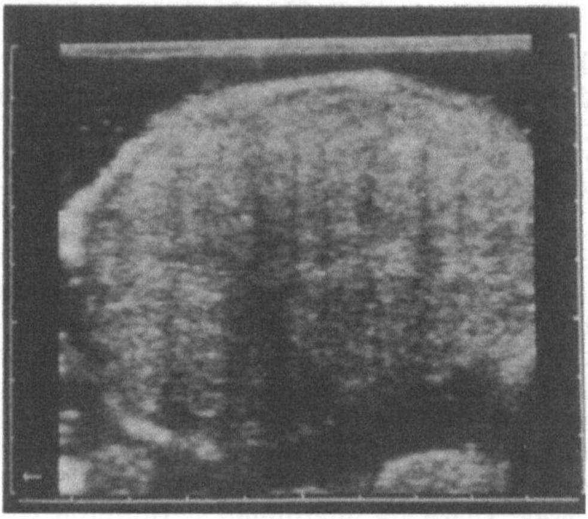

Fig. 22: Ovary of a mare with a granulosa-cell tumour of the sertoli-cell type. The section shows a relatively homogeneous echogenity and no internal divisions. Narrow non-echogenic bands arising from small circumscribed areas of ossification can be followed into the depth of the structure.

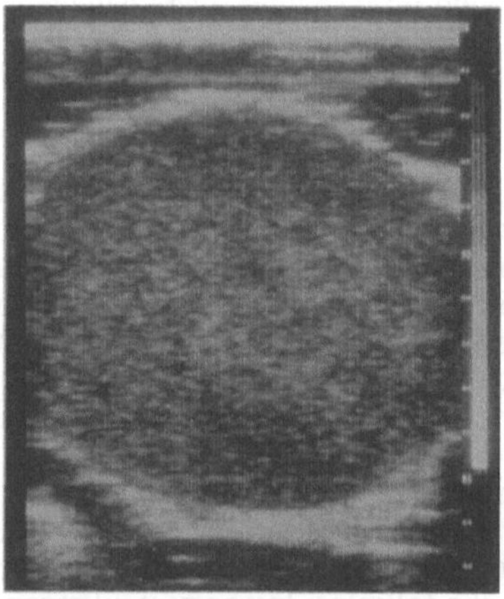

Fig. 23: Ovary of a mare with a singular cyst 10 cm in diameter and of average echogenity. The content was sero-haematic. Histological examination of the wall showed neoplastic changes in the form of a granulosa-cell tumour.

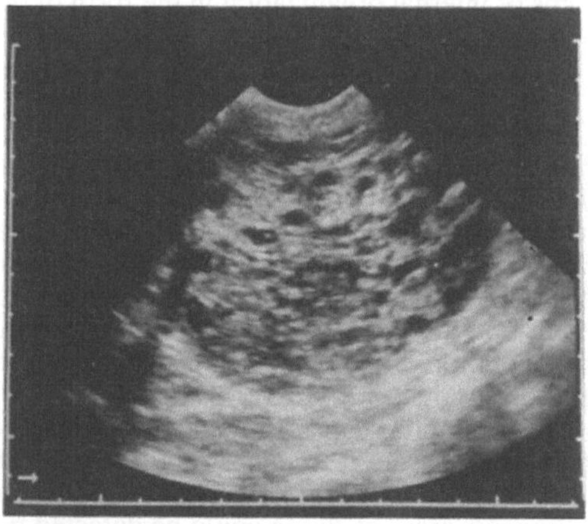

Fig. 24: Ovarian cystadenoma in a mare characterized by numerous small cysts embedded in a relatively large amount of echogenic neoplastic tissue.

vesicular structures than can be counted on normal ovaries with follicles (Fig. 21). An ovary usually only presents 5-10 follicles larger than 10 mm in diameter during the oestrus cycle. It has been demonstrated that there are many more cystic bodies present in the case of ovarian tumours. Ultrasonic studies performed so far, as well as the examinations of excised ovaries, have revealed up to 50-60 such vesicular structures on a single ovary.

Granulosa cell tumours were shown to have cysts with a diameter ranging from a few mm to many centimeter. An intact ovary is characterized by the simultaneous presence of follicles in different stages of development and therefore of various sizes. Thus, it is exceptional to find more than 2 or 3 cysts with a diameter of 30-50 mm, the majority of the cysts being 5-20 mm in diameter. In contrast to this, ultrasonic images of granulosa-cell tumours may either reveal a large number of extremely big cysts, or, on the other hand, numerous very small cysts closely packed on the ovary.

The internal echogenity of follicular structures of granulosa-cell tumours varies greatly. Sometimes vesicles with non-echogenic cavities and others with echo-rich cavities are found side by side on the same ovary. In the majority of the granulosa-cell tumours investigated so far, the total cross-sectional area of cystic structures was greater than the area of echo-rich neoplastic tissue.

Besides these multicystic granulosa-cell tumours, one tumour with relatively compact tissue and no detectable chambers was found (Fig. 22). The sonographic cross-section revealed a more or less homogeneous echogenity with no vesicular structures. Highly echogenic circumscribed areas representing centers of ossification were found scattered throughout the tumour. After exenteration of the ovary, the firm consistency of the tumour was verified. Histologically, it was found to be a granulosa-cell tumour of the sertoli cell type.

In another case of an ovarian tumour, the sonographic image closely resembled that of an ovarian haematoma (Fig. 23). The ovary consisted almost exclusively of a single, 10 cm large follicle with average echogenity. Study of the excised organ showed the follicle to be filled with sero-haematic fluid. Neoplastic tissue of the granulosa-cell type was found in the follicular wall on histological examination. Epithelial cell proliferations showing sertoli cell-like formations projected into the cavity.

Ovarian neoplasms in the form of cystadenomas are seen less frequently in the mare than granulosa-cell tumours. The neoplastic tissue of cystadenomas investigated ultrasonically, showed a mixture of various echogenic intensities. Numerous cystic structures were disseminated in the tissue (Fig. 24). These echo-weak vesicular bodies had diameters of only a few millimeters. The proportion of echogenic neoplastic tissue was greater than the nonechogenic cystic areas.

In general different ovarian tumours can have a characteristic sonographic pattern but it is not possible to diagnose accurately the type of the toumor using ultrasound scanning.

Conclusions

The availability of ultrasound instrumentation has opened new possibilities in the diagnosis of pathological phenomena of the reproductive tract of the mare. Providing a noninvasive form of visual access to the uterus and ovaries, it facilitates a more prompt, objective assessment of changes than was possible by rectal palpation.

Aspects of a reliable diagnosis of endometrial cysts and their differentiation from an early embryonic vesicle are discussed. Ultrasound technology provides the first non-invasive, safe method for evaluating the vitality and the developmental stage of an embryo or foetus. It is demonstrated how complications in pregnancy leading to early embryonic death or abortion can be diagnosed very early and followed up. Ultrasonography is a useful method for the control of normal and disturbed uterine involution after parturition.

Fluid collections in the uterus during dioestrus have a pathological significance whereas, if they occur during oestrus, subsequent pregnancy cannot be excluded.

Ultrasonographic techniques can also be successfully applied to the diagnosis and evaluation of pathological conditions of the ovaries. An exact differentiation of anovulatory follicles was not always possible in the past, using rectal examination procedures. Absence of ovulation and the following development of persisting follicles can be determined and investigated by sonographic examinations. Those structures either become luteinized or haemorrhagic. Recent work has established parameters useful for an efficient ultrasound diagnosis and differentiation of ovarian neoplasms.

Since the initial reports on its efficiency, ultrasound has become the technique of choice for the clinician and for research purposes dealing with the reproductive tract of the mare.

References

Adams, G. P., Kastelic, J. P., Bergfelt, D. R. and Ginther, O. J. (1987) 'Effect of uterine inflammation and ultrasonically-detected uterine pathology on fertility in the mare', J. Reprod. Fert., Suppl. 35, 445-454.

Colles, C. M. and Parkes, R. D. (1978): Foetal electrocardiography in the mare, Equine vet. J. 10, 32-37.

Darenius, K., Kindahl, H. and Madej, A. (1988) 'Clinical and endocrine studies in mares with known history of repeated conceptus losses', Theriogenology 29, 1215-1232.

Ginther, O. J. and Pierson, R. A. (1984 a) 'Ultrasonic anatomy of equine ovaries', Theriogenology 21, 471-483.

Ginther, O. J. and Pierson, R. A. (1984 b) 'Ultrasonic anatomy and pathology of the equine uterus', Theriogenology 21, 505-516.

Ginther, O. J., Bergfelt, D. R., Leith, G. S. and Scraba, S. T. (1985) 'Embryonic loss in mares: incidence and ultrasonic morphology', Theriogenology 24, 73-86.

Kähn, W. und Leidl, W. (1987 a) 'Echographische Befunde an Ovarien von Stuten', Tierärztl. Umsch. 42, 257-266.

Kähn, W. und Leidl, W. (1987 b) 'Die Ultraschall-Biometrie von Pferdefeten in utero und die sonographische Darstellung ihrer Organe', Dtsch. tierärztl. Wschr. 94, 509-515.

Kaspar, B., Kähn, W., Laging, C. und Leidl, W. (1987) 'Endometriumzysten bei Stuten. Teil 1. Post-mortem-Untersuchungen: Vorkommen und Morphologie', Tierärztl. Prax. 15, 161-166.

Kenney, R. M. and Ganjam, V. K. (1975) 'Selected pathological changes of the mare uterus and ovary', J. Reprod. Fert., Suppl. 23, 335-339.

Leidl, W., Kaspar, B. und Kähn, W. (1987) 'Endometriumzysten bei Stuten. Teil 2. Klinische Untersuchungen: Vorkommen und Bedeutung', Tierärztl. Prax. 15, 281-289.

Leidl, W. und Kähn, W. (1984) 'Differentialdiagnostische Befunde bei der Frühträchtigkeitsuntersuchung von Stuten mit dem Ultraschallverfahren (Echographie)', Vlaams diergeneesk. Tijdschr. 53, 170-179.

Leith, G. S. and Ginther, O. J. (1985) 'Mobility of the conceptus and uterine contractions in the mare', Theriogenology 24, 701-711.

Matsui, K., Sugano, S. and Masuyama, I. (1985) 'Changes in the fetal heart rate of thoroughbred horse through the gestation', Jap. J. vet. Sci. 47, 597-601.

McKinnon, A. O., Squires, E. L., Harrison, L. A., Blach, E. L. and Shideler, R. K. (1988) 'Ultrasonographic studies on the reproductive tract of mares after parturition: Effect of involution and uterine fluid on pregnancy rates in mares with normal and delayed first postpartum ovulatory cycles', J. Am. vet. med. Ass. 192, 350-353.

Palmer, E. and Driancourt, M. A. (1980) 'Use of ultrasonic echography in equine gyne-

cology' Theriogenology 13, 203-216.
Pierson, R. A. and Ginther, O. J. (1985) 'Ultrasonic evaluation of the corpus luteum in the mare', Theriogenology 23, 795-806.
Pipers, F. S. and Adams-Brendemuehl, C. S. (1984) 'Techniques and applications of transabdominal ultrasonography in the pregnant mare', J. Am vet. med. Ass. 185, 766-771.
Simpson, D. J., Greenwood, R. E. S., Ricketts, S. W., Rossdale, P. D., Sanderson, M. and Allen, W. R. (1982) 'Use of ultrasound echography for early diagnosis of single and twin pregnancy in the mare' J. Reprod. Fert., Suppl. 32, 431-439.
Squires, E. L., McKinnon, A. O. and Shideler, R. K. (1988) 'Use of ultrasonography in reproductive management of mares', Theriogenology 29, 55-70.
Staudach, A. (1986) Fetale Anatomie im Ultraschall, Springer-Verlag, Berlin, Heidelberg, New York, London, Paris, Tokyo.
Valon, F., Sedard, F. et Chaffaux, St. (1982) 'Echotomographie en temps réel de l'utérus chez la jument', Bull. Acad. vét. Fr. 55, 187-211.
White, R. A. S. and Allen, W. R. (1985) 'Use of ultrasound echography for the differential diagnosis of a granulosa cell tumour in a mare' Equine vet. J. 17, 401-402.
Will, K., Kähn, W. und Leidl, W. (1988) 'Sonographische Untersuchungen über die präovulatorische Follikelentwicklung bei der Stute' Dtsch. tierärztl. Wschr. 95, 362-365.

ULTRASONIC CHARACTERISTICS OF PHYSIOLOGICAL STRUCTURES ON BOVINE OVARIES

M.C. Pieterse
Department of Herd Health and Reproduction
Faculty Veterinary Medicine
Utrecht, the Netherlands.

INTRODUCTION

The use of diagnostic ultrasonography in veterinary medicine in general (23), and especially for the assessment of the bovine genital tract and ovarian structures (7,6,13,22), is not yet widespread, but is increasing rapidly. The topography of the bovine genital tract which cannot be visualized as easily as, for example, the genital tract of the mare, is the main reason why this technique is not yet so widespread in cows. The ovaries and the tips of the bovine uterine horns are not always easy to reach with the ultrasound transducer via the rectum, vagina or via the abdominal wall. Ultrasonography may be necessary when rectal palpation of the genital tract as well as hormone analysis do not give the desired information about normal or pathological ovarian activity. Visualization of the ovaries via rectum or vagina can give us extra information about the follicle population, the presence of functional corpora lutea, follicular and luteinized cysts, haematomas in the ovaries, ovarian responses to hormonal treatment (1,25) and abnormalities in the ovarian region. Other applications of ultrasonography in cows besides the study of ovarian structures include early pregnancy diagnosis (1,4,7,12,16,24,26), check of the age, sex and/or viability of the fetus, and aspiration of oocytes, the so-called "ovum pick-up" (15). The ultrasound scanning of bovine ovaries and the diagnostic reliability of this technique was first reported by Pierson et al. (11,13) and Quirk et al. (18). Changes in growth patterns of the follicles in the ovaries with the aid of (daily) ultrasound scanning have also been studied (22).
A correct interpretation of the common physiological structures in the ovaries visualized by ultrasonography is of primary importance for those people who wish to make use of this modern technique in bovine gynaecology.
This paper describes the main characteristics of the common bovine ovarian structures throughout the oestrous cycle visualized by means of trans-rectal and/or trans-vaginal ultrasonography.

M. M. Taverne and A. H. Willemse (eds.), Diagnostic Ultrasound and Animal Reproduction, 37–51.

PREPARATION OF THE ANIMAL

Restraining the animal in a standing position facilitates the scanning. Generally no special treatment is required for a short scan of the uterus and ovaries. However, epidural anaesthesia with 2.5 cc Lidocaine is sometimes necessary to prevent abdominal straining which may interfere with scanning and manipulation of the ovaries and uterus. A more thorough examination of the uterus and ovaries requires sedation of the animal, combined with a relaxation of the intestines. This may be achieved with 0.1cc Domosedan^{R} (detomidine hydrochloride 10 mg/ml, Smith Kline Animal Health Ltd.) per 100 kg. The direct result of an intravenous injection is an adequate relaxation of the intestines and a quiet, standing animal. Faeces should always be removed from the rectum before the start of trans-rectal or trans-vaginal ultrasonography.

ULTRASOUND EQUIPMENT

A real time, B-mode diagnostic ultrasound scanner equipped with a linear array or a sector scan transducer of at least 5 mHz is necessary for good resolution with clear images of the structures. A mark or groove on the grip of the transducer, or on the probe itself, is necessary for orientation of the scanning direction. Bright sunlight near or on the monitor makes the images hard to interpret, so scanning has to be done in a more or less dark environment.
Scanning of the ovaries can be done:
a. via the rectum (20)
b. via the vagina (15)

Ad a: trans-rectal ultrasonography

The trans-rectal scan is preferred to the trans-vaginal scan, even if both approaches lead to the same result. Trans-vaginal scanning requires additional hygienic measures to prevent contamination of the vagina. The trans-rectal scan is usually performed with a linear-array transducer, although new trans-rectal sector transducers have been developed. The smaller the transducer the easier the intrarectal manipulation will be. A disadvantage of this method is that scanning of the ovaries via the rectum can only be done with one hand.
The ovaries may be located by following the uterine horns. Hand and transducer are covered with contact-gel. Both are brought to a position dorsal of the ovary. The ovary is monitored by moving the transducer slowly from left to right and vice versa. It is sometimes necessary to fix the ovary (if possible) during scanning between the ring and middle-finger, while the other fingers fix the transducer in the

right place. Retraction and reflection of the uterus and ovaries (14) may facilitate the procedure.

Ad b: trans-vaginal ultrasonography

The trans-vaginal scan is used when more detailed information about the structures in the ovaries is required. Closer contact between transducer and ovary and a better positioning of the ovary against the transducer will give a better resolution of the structures. Also a more systematic scan of the ovary is possible, because one hand manipulates the ovary, the other the transducer. A transvaginal scan of the ovaries may be done with either a sector transducer or a linear array transducer. For both transducers an elongated grip is needed to guide the transducer within the vagina from the outside (Fig. 1).

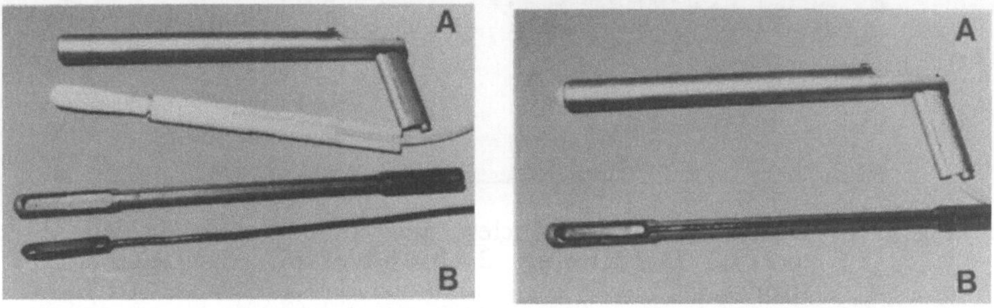

Fig. 1. The figure on the left shows a disassembled sector transducer (A) and a linear array transducer (B) with their grips; the right figure shows the assembled equipment ready for use.

Scanning of the ovaries via the vagina has to be done with both hands. It makes a more systematic scan of the ovary possible, because one hand manipulates the ovary, the other the transducer. The ovary is retracted, via the rectum, to a position caudal to the cervix and placed against the left or right side of the vaginal wall; the other hand places the transducer, via the vagina, against the ovary.
Both the grip of the transducer and the vulva and perineum have to be cleaned thoroughly before trans-vaginal ultrasonography.

OVARIAN STRUCTURES

Follicles

Follicle growth takes place during the whole oestrous cycle. Follicles of different sizes (Fig. 2) may be visualized at

any time during the oestrous cycle, but the smaller follicles (2-5 mm) dominate during the first days after ovulation (9).
Follicles are easy to distinguish using ultrasonography. They appear as black, roughly circumscribed areas on the ultrasound images. Normal follicles measure up to 25 mm. Visualization of follicles as small as 3-4 mm is possible (11) with a 5 mHz transducer. It is generally believed that the pre-ovulatory follicle is "selected" from a pool of similarly sized follicles. It ensures its "dominance" over the other follicles in the ovary by, in some way, actively suppressing their further development (3).

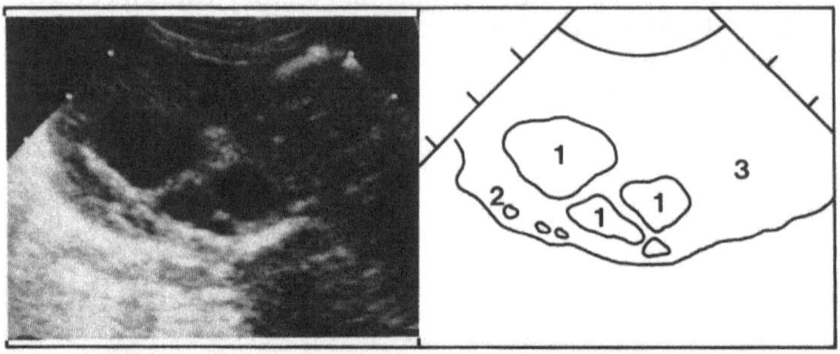

Fig. 2. Different sized follicles and bloodvessels in an ovary. 1 = follicle, 2 = bloodvessel, 3 = ovarian stroma.

The pre-ovulatory follicle can, in most cases, be observed one to three days prior to oestrus (3). Before this time the pre-ovulatory follicle is not consistently the largest follicle (2).
Ovulation may be detected by the disappearance of the pre-ovulatory follicle or by the strongly diminished size of this follicle. A corpus haemorrhagicum (young corpus luteum) is recognized on the monitor as a less defined black image with echogenic spots. Young corpora lutea are always hard to distinguish; especially when the location of the preovulatory follicle was not visualized during the oestrous period, before the ovulation took place.
Apposed walls of similarly sized follicles are often straight; a small follicle adjacent to a larger follicle frequently appears as a convex impingement on the surface of the larger follicle (Fig. 3).
Blood vessels (Figs. 2 and 9) may be confused with small follicles. Extended bloodvessels are seen during the luteal phase as a result of an increased blood flow to the corpus luteum (10). They can be identified by moving the transducer: bloodvessels can, unlike follicles, be traced during scanning over a longer distance and are generally seen on

the border of the ovary, in the ovarian hilus and around the corpus luteum.

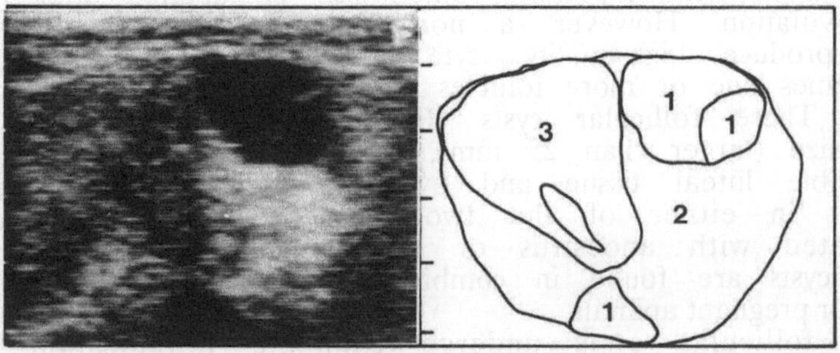

Fig. 3. Impingement of a small follicle into a larger follicle. 1 = follicle, 2 = ovarian stroma, 3 = corpus luteum.

In an embryo transfer program the ovaries are stimulated with pregnant mare serum gonadotrophin (PMSG) or follicle stimulating hormone (FSH). The echoscopic image of those ovaries should show an increased number of enlarged pre-ovulatory follicles (Fig. 4).

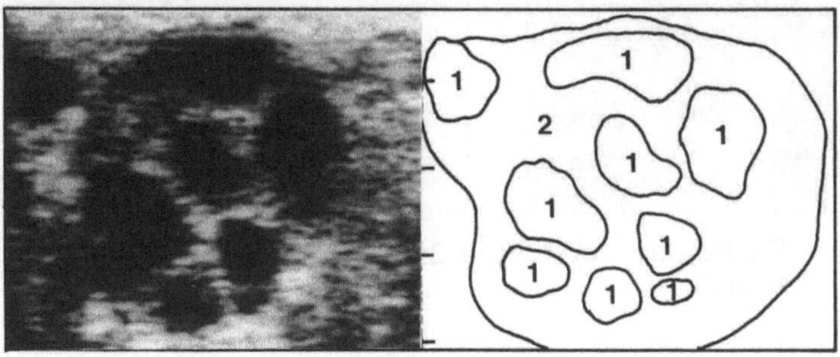

Fig. 4. Multifollicular ovary after stimulation with pregnant mare serum gonadotrophin (PMSG). 1 = follicle, 2 = ovarian stroma.

Rectal palpation gives only an impression of the enlargement of the ovaries after stimulation with hormones, but does not give a proper indication of the number and size of the developing (pre-ovulatory) follicles. A palpated, enlarged, ovary indicates merely that the animal has reacted to the hormonal treatment. A differentiation between the presence of several normally sized ($\geqslant 25$ mm) follicles, which will probably ovulate, and enlarged cystic follicles, which will probably not ovulate, cannot always be made by rectal palpation. However, this may be done using ultrasonography.

It defines for example if it will be worthwhile to continue an embryo transfer program, especially when artificial insemination with expensive semen will be applied after the superovulation. However, a normal sized follicle does not always produce a high-quality oocyte.
Sometimes one or more follicles continue to grow and become cystic. These follicular cysts (Fig. 5) are characterized by their size (larger than 25 mm), the absence of echoscopically visible luteal tissue and by the absence of a corpus luteum in either of the two ovaries. Usually they are associated with anoestrus or nymphomania. In exceptional cases cysts are found in combination with corpora lutea in cyclic or pregnant animals.
Some follicular cysts undergo complete luteinization or luteinize partly. Both structures cause anoestrus. A luteinized follicular cyst will be monitored as an (unruptured) follicle (black on the screen) lined by an (irregular) enlarged echogenic layer (grey on the screen), supposed to be luteinized cells (Fig. 6).

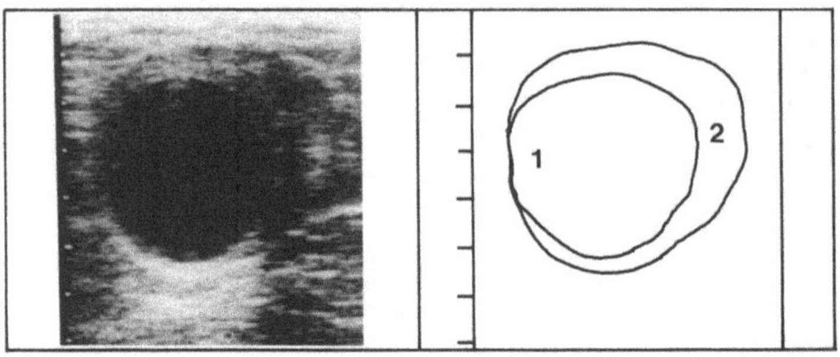

Fig. 5. Follicular cyst (⩾ 25mm). 1 = cyst, 2 = ovarian stroma.

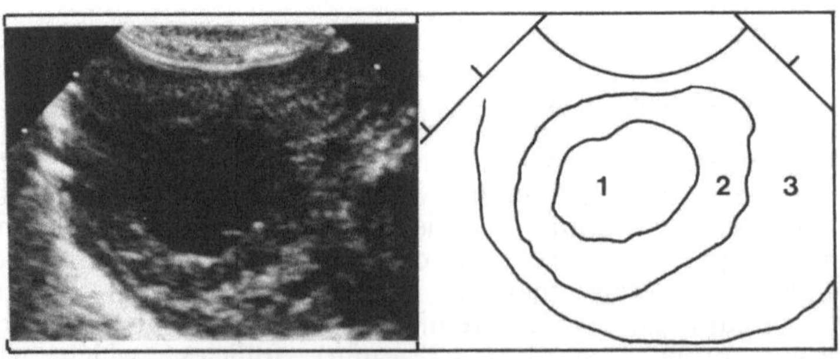

Fig. 6. A luteinized follicular cyst. 1 = cyst, 2 = luteinized cells, 3 = ovarian stroma.

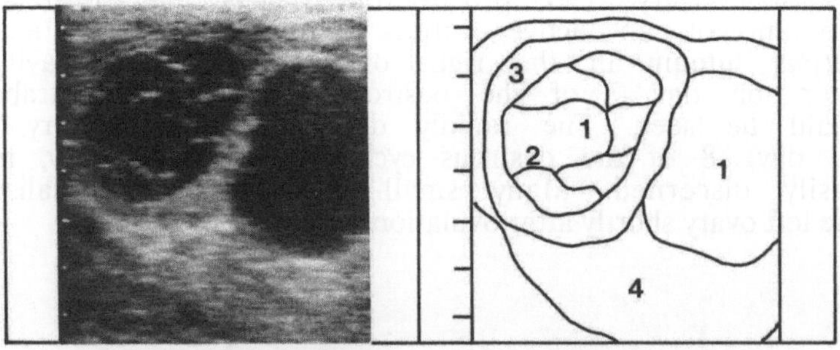

Fig. 7. Luteinized follicular cyst with trabeculae and a lining of luteal granulosa cells. 1 = cyst, 2 = trabeculae, 3 = luteinized cells, 4 = ovarian stroma.

A luteinized follicular cyst should not be confused with a corpus luteum with a cavity (Fig. 10). Differentiation by means of rectal palpation between follicular cysts, luteinized follicular cysts and corpora lutea with cavities is not easy. Ultrasonography helps to differentiate between these structures. A study of the accuracy of the diagnosis made using ultrasonography will be published in a later study (17), in which a number of ovaries were dissected after rectal palpation and ultrasonography.

In some cases the layer of luteinized cells divides the cyst in a network of compartments which can clearly be seen on an echoscopic image (Fig. 7).

The growth and regression of different follicles during a complete oestrous cycle is a rather complicated process. Repeated ultrasonography is necessary for the correct interpretation of these dynamics. Longitudinal (daily) echoscopic studies provide a better insight into the pattern of follicle growth in cows. Various authors, such as Rajakoski (19), Savio (21), and Sirois (22), published theories about the existence of growth waves of follicles during the oestrous cycle. Daily echoscopic observation is not invasive for the animals, and provides us with much information about the follicular growth and the dynamics of the ovaries. However, the complexity of follicular growth is demonstrated by the diversity of opinion expressed in these publications. It is generally accepted that three waves of follicular growth occur during one oestrous cycle. The growth of follicles and corpora lutea during a complete oestrous cycle has been made echoscopically visible in one of our studies in heifers with regular oestrous cycles. An example of an oestrous cycle of one heifer is shown schematically in figure 8. Initially this animal ovulated from the right ovary, but during the next oestrus, three weeks later, from the left ovary. Her progesterone levels indicate the development of an active corpus luteum after ovulation. The corpus luteum in the right ovary developed a cavity, and later, on day 7 of the oestrous cycle, a small trabeculum could be seen. The rapidly developing pre-ovulatory follicle on day 18 of the oestrous cycle, on the left ovary, may be easily discerned. Many small follicles were visualized on the left ovary shortly after ovulation.

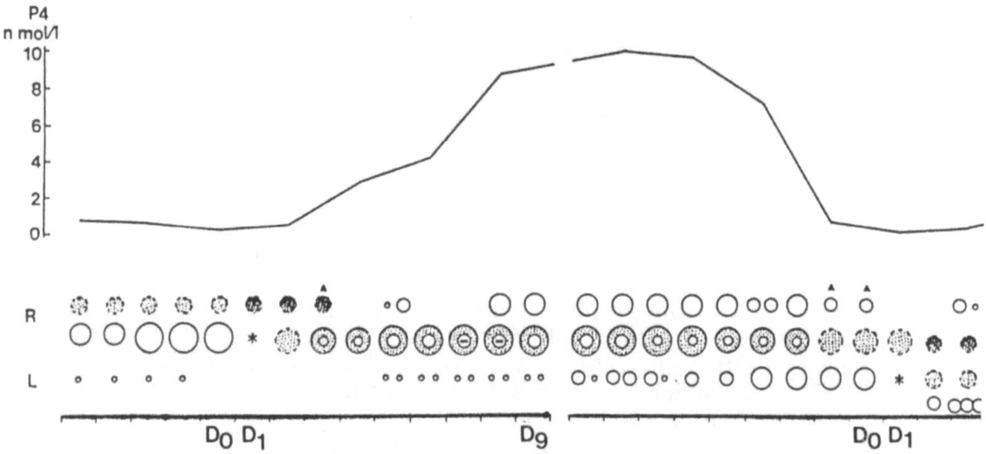

Fig. 8. Schematic representation of the dynamics in the two ovaries of a heifer during one oestrous cycle observed with a 5 mHz linear array trans-rectal transducer. The blood-progesterone levels during this oestrous cycle are given in nmol/l. Day 9, 10, 11, and 12 did not show differences during the study.

D0 = oestrus
P4 = blood-progesterone level nmol/l
L = left ovary
R = right ovary
▲ = bloodvessels near the ovary
* = ovulated follicle
∘ = follicle ⩽ 6 mm
○ = follicle > 6 mm and ⩽ 12 mm
◯ = follicle > 12 mm and ⩽ 18 mm
(symbol) = corpus luteum
(symbol) = corpus luteum with cavity and trabecula
(symbol) = corpus luteum with cavity

Corpora lutea

Corpora lutea (Figs. 9, 10 and 11) have, in contrast to the liquid-filled follicles, a granular echogenic structure which intensifies during the luteal phase. The image is grey, whereas the images of follicles (and blood vessels) are black and that of the ovarian stroma is diffuse white. A boundary may be seen as a result of this contrast between the corpus luteum and the stroma.

Young corpora lutea (Fig. 9) appear as a light grey echogenic structures two to four days after ovulation. They are sometimes hard to discern in the ovarian stroma, even when a 7.5 mHz trans-vaginal scanner is used. In some cases the old corpus luteum of a former oestrous cycle (corpus luteum atreticum) may be visualized alongside the young developing corpus luteum (Fig. 8). In general, old corpora lutea of earlier oestrous cycles are not discernable, as they have the same echogenicity as the stroma. The diameter of an active corpus luteum can increase up to approximately 30 mm. At the end of the luteal phase, about day 17 of the oestrous cycle, the corpus luteum begins to decrease in size (6).

Several corpora lutea may be visualized on both ovaries some days after ovulation when a successful PMSG-stimulation is applied during an embryo transfer program. A poor reaction results in too many persisting follicles or an absence of developing corpora lutea. The number of visualized corpora lutea gives an impression of the, maximum number of embryos, which may be obtained after flushing of the uterus one week, or twelve days, after insemination. Counting the number of corpora lutea by rectal palpation is much more difficult.

Two structures are sometimes seen during scanning in the corpora lutea:

-a liquid-filled central cavity (Fig. 10) and

-a white thin line (trabeculum) in the centre of the corpus luteum (Fig. 11).

The liquid-filled, round to oval central cavity results in a black image and is regularly seen (5,6,8). Kito et al. (8) observed during their study of cows with a normal oestrous cycle a cavity in 37.2% of the corpora lutea. These structures are sometimes referred to as cystic corpora lutea or corpora lutea cysts. In our opinion this is incorrect. It has been demonstrated (8) that these corpora lutea with a liquid-filled central cavity are normally functioning corpora lutea. They produce the same progesterone levels in periferal blood as corpora lutea without a central cavity. Some cavities remain throughout the entire oestrous cycle, while others disappear.The cavity in the corpus luteum should not be confused with a luteinized follicular cyst (Fig. 7), which is a pathological degeneration of the persisting follicle. The luteinized follicular cysts have, in contrast to the corpora lutea, an irregular and thin

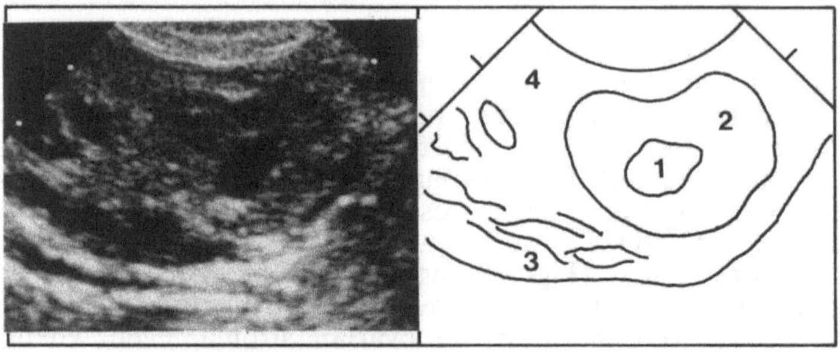

Fig. 9. Young corpus luteum with a cavity. Blood vessels are clearly seen on the border of the ovary. 1 = cavity, 2 = luteal cells, 3 = bloodvessels, 4 = ovarian stroma.

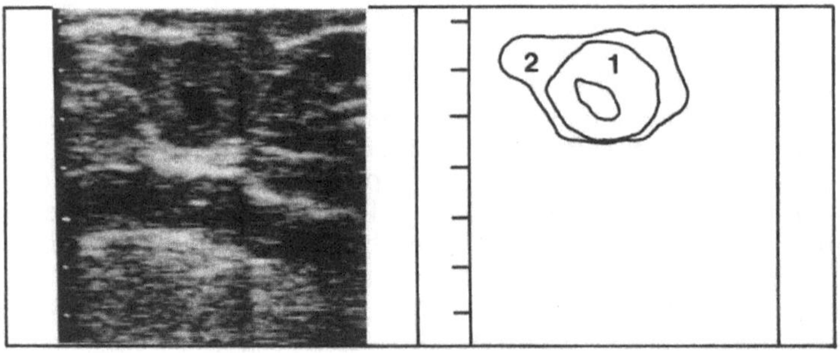

Fig. 10. An active, normal corpus luteum with a central cavity. 1 = corpus luteum, 2 = ovarian stroma.

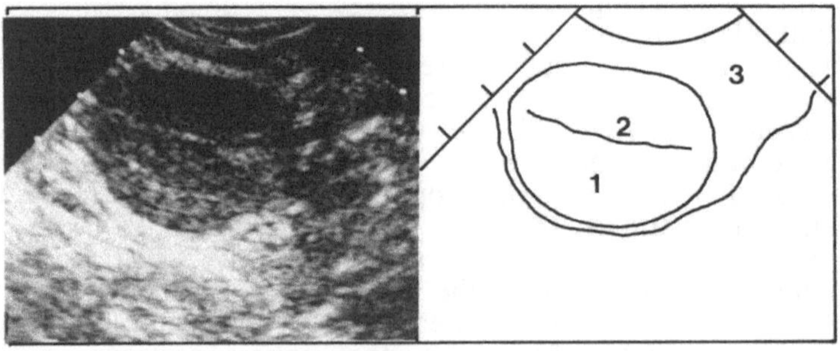

Fig. 11. An active, normal corpus luteum with a trabeculum. 1 = corpus luteum, 2 = trabeculum, 3 = ovarian stroma.

lining of luteal tissue, and cavities with diameters exceeding 25 mm. The trabeculum, seen regularly in the mid-cycle corpus luteum, is probably the central part of connective tissue in the corpus luteum.
It is never marked (17) as a special structure after dissection of these corpora lutea. This so-called trabeculum of connective tissue is only distinguishable ultrasono-graphically in functional corpora lutea and it provides the vascularisation in the luteal tissue of a corpus luteum (8).
The trabeculum, as well as the liquid-filled cavity, are generally located in the middle of the corpus luteum.

CONCLUSION

The literature of the past years shows that ultrasonography has proved to be a useful technique for both diagnostic and experimental purposes in bovine gynaecology.
In general, a veterinary practitioner will not use ultrasonography if rectal palpation of the genital tract provides him with sufficient information about the ovarian activity or about pregnancy. Rectal palpation is easier and quicker to execute than ultrasonography and requires no expensive equipment. However, ultrasonography may provide additional information about the palpated structures. The ovaries of animals with an enlarged uterus (late pregnancy, puerperium, pyometra) or adhesions are often difficult to manipulate or visualize.
In our opinion the best images are obtained by using the trans-vaginal technique. This method makes it possible to perform a more systematic scan of the ovaries. Unlike during rectal scanning, the faeces will not distort the image. Smaller structures, such as follicles smaller than 5 mm, and corpora lutea up to day 4 of the oestrous cycle, are easier to detect, although this has still to be proved by comparative studies. A disadvantage of the vaginal scan is the manipulation in the vagina, which requires thorough cleaning of the vulva and perineum in order to prevent faecal contamination of transducer and genital tract. Although a better image is obtained with a trans-vaginal scan, it should only be applied when rectal palpation and a trans-rectal scan does not give conclusive information. Until now, no extensive studies have been made which compare rectal palpation with ultrasonography of the ovaries. In one of our recent studies (17) the reliability of ultra-sonography for the interpretation of the different structures on bovine ovaries has been investigated. The ovaries were first palpated and subsequently scanned; the results were verified by dissection of the ovaries after slaughtering on the same day. It showed that far more information is obtained using ultrasonography than by rectal palpation. This means that ultrasonography can give the veterinary surgeon additional informa-

tion about the ovarian structures, sometimes necessary for a good diagnosis and treatment. Small follicles cannot be palpated, but are of minor importance to the general practitioner. However, they are of great importance for "ovum pick-up" combined with in vitro maturation/fertilization programs. They are also important when stimulation with PMSG or FSH is applied for super-ovulation and embryo transfer. These small follicles can be easily scanned using a 5 mHz, or, even better, a 7.5 mHz transducer. The 7.5 mHz transducer (rectal or vaginal) has a higher resolution and is therefore preferred for visualization of the other ovarian structures too.

Very young corpora lutea or corpora haemoragica are hard to palpate and may easily be confused with pre-ovulatory follicles. In such cases ultrasonography is necessary, especially when advice about insemination is requested. The young corpora lutea are not easy to discern using ultrasonography, but pre-ovulatory follicles are. The presence of a pre-ovulatory follicle justifies insemination; the disappearance of the same folicle is the first indication that ovulation took place. Active corpora lutea are usually easy to palpate and visualize using ultrasound. An estimation of the exact day of the oestrous cycle cannot be made with ultrasound on the basis of the images of the corpus luteum on the monitor. A differentiation can only be made between a functional and non-functional corpus luteum. This makes it possible to differentiate between the beginning, middle and end of the oestrous cycle. A cavity in a corpus luteum, which is not pathological and may as well occur in pregnant cows, may be mistaken for a cyst during rectal palpation; this mistake cannot easily be made using ultrasonography, where the thickness of luteal tissue and the diameter of the cavity/cyst may be visua-lized. Corpora lutea have a thick layer of luteal tissue and a relatively small central cavity, unlike cysts. This know-ledge is essential for the differentiation between follicle cysts and luteinized follicle cysts when a special therapy has to be applied. For example, prostaglandins can be applied when luteal tissue is present, whereas releasing hormones or luteinizing hormones are applied when luteal tissue is lacking. A few days after treatment with prosta-glandins it is possible to palpate the regression of the corpus luteum and, to a lesser extent, the regression of luteal tissue in luteinized follicle cysts. However, this regression is usually accompanied by (visible) oestrous symptoms. The result of a releasing hormone or luteinizing hormone therapy can only be verified by blood/milk progesterone analysis or ultrasonography.

The many advantages and uses of ultrasonography in bovine gynaecology increase with the experience of the veterinary surgeon. In the near future ultrasonography will prove to be an indispensable technique in modern bovine gynaecology under field conditions as well as in clinics and research centres. More information about follicle dynamics, the influence of various drugs on the follicle population, and ovarian pathology may be obtained in the near future with the aid of ultrasonography. However, a basic knowledge of the normal physiological structures will remain the essential basis for an optimal use of this new and fascinating technique.

LITERATURE

1. Chupin, D., Procureur, R. (1983) 'Prediction of bovine ovarian response to PMSG by ultrasonic echography', Theriogenology, 19, 1, 119.

2. Dufour, J., Whitmore, H.L., Ginther, O.J., Casida, L.E. (1971) 'Identification of the ovulating follicle by its size on different days of the oestrous cycle in heifers', Journal of Animal Science, 34, 85-87.

3. Fortune, J.E., Sirois, J., Quirk, S.M. (1988) 'The growth and differentiation of ovarian follicles during the bovine estrous cycle', Theriogenology, 29, 1, 95-109.

4. Hansen, C. and Delsaux, B. (1987) 'Use of transrectal B-mode ultrasound imaging in bovine pregnancy diagnosis', Veterinary Record 121, 200-202.

5. Horstmann, G., Schwarz, R., Neurand, K. (1973) 'Die Corpus Luteum Zyste des Rindes; Mikromorphologie und Diskussion ihre Entstehung', Zbl. Vet. Med. A, 20, 292.

6. Kähn, W., Leidl, W. (1986) 'Die Anwendung der Echographie zur Diagnose der Ovarfunktion beim Rind', Tierärztliche Umschau, 41, 1, 3-12.

7. Kastelic, J.P., Curran, S., Pierson, R.A., Ginther, O.J. (1988) 'Ultrasonic evaluation of the bovine conceptus', Theriogenology 29, 39-54.

8. Kito, S., Okuda, K., Miyazawa, K., Sata, K. (1986) 'Study on the appearance of the cavity in the corpus luteum of cows by using ultrasound scanning', Theriogenology, 25, 325-333.

9. Kruip, Th.A.M. (1982) 'Macroscopic identification of tertiairy follicles >2mm in the ovaries of cycling cows', In: Karg, H. and Schallenberg, E. (eds). Factors influencing fertility in the post partum cow. Current topics in Veterinary Medicine and Animal Science. Martinus Nijhoff Pub., the Hague, 95-101.

10. Niswender, G.D., Reimers, T.J., Diekman, M.A., Nett, T.M. (1976) 'Blood flow: A mediator of ovarian function', Biology of Reproduction, 14, 64-81.

11. Pierson, R.A., Ginther, O.J. (1984) 'Ultrasonography of the bovine ovary', Theriogenology, 21, 3, 495-504.

12. Pierson, R.A., Ginther, O.J. (1984) 'Ultrasonography for detection of pregnancy and study of embryonic development in heifers', Theriogenology, 22, 2, 225-233.

13. Pierson, R.A., Ginther, O.J. (1987) 'Reliability of diagnostic ultrasonography for identification and measurement of follicles and detecting the corpus luteum in heifers', Theriogenology, 28, 6, 929-936.

14. Pieterse, M.C., Willemse, A.H. (1983) 'Rectal palpation in the diagnosis of pregnancy in cows', Pro Veterinario, 2, 5-8.

15. Pieterse, M.C., Kappen, K.A., Kruip, Th.A.M., Taverne, M.A.M. (1988) 'Aspiration of bovine oocytes during transvaginal ultrasound scanning of the ovaries'. Theriogenology, 30, 4, 751-762.

16. Pieterse, M.C., Scenzi, O., Willemse, A.H., Csaba, B., Taverne, M.A.M. 'Early pregnancy diagnosis in cattle by means of linear-array real-time ultrasound scanning of the uterus and a qualitative and quantitative milk progesterone test', (Submitted for publication).

17. Pieterse, M.C., Taverne, M.A.M., Kruip, Th.A.M. 'Evaluation of the bovine ovary with rectal palpation, echoscopy and ovariectomy', (Submitted for publication).

18. Quirk, S.M., Hickey, G.J., Fortune, J.E. (1986) 'Growth and regression of ovarian follicles during the follicular phase of the oestrous cycle in heifers undergoing spontaneous and PGF2ã-induced luteolysis', Journal of Reproduction and Fertility, 77, 211-219.

19. Rajakoski, E. (1960) 'The ovarian follicular system in sexually mature heifers with special reference to seasonal, cyclical and and left right variations', Acta Endocrinologica, suppl. 52, 34, 7-68.

20. Reeves, J.J., Rantanen, N,W., Hauser, M. (1984) 'Transrectal real-time ultrasound scanning of the cow reproductive tract', Theriogenology, 21, 3, 485-494.

21. Savio, J.D., Keenan, L., Boland, M.P., Roche, J.F. (1988) 'Pattern of growth dominant follicles during the oestrous cycle of heifers', Journal of Reproduction and Fertility, 83, 1-9.

22. Sirois, J and Fortune, J.E. (1988) 'Ovarian follicular dynamics during the oestrous cycle in heifers monitored by real-time ultrasonography'. Biology of Reproduction, 39, 308-317.

23. Taverne, M.A.M. (1984) 'The use of linear-array real-time echography in veterinary obstetrics and gynaecology', Tijdschrift voor Diergeneeskunde, 109, 12, 494-506.

24. Taverne, M.A.M., Szenci, O., Szetag, J., Piros, A. (1985) 'Pregnancy diagnosis in cows with linear-array real-time ultrasound scanning: a preliminary note', Veterinary Quarterly, 7, 4, 264-270.

25. Thayer, K.M., Forest, D.W., Welsh, T.H., Jr. (1985) 'Real-time ultrasound evaluation of follicular development insuperovulated cows', Theriogenology, 23, 1, 233.

26. White, J.R., Russel, A.J.F., Wright, J.A., Whyte, T.K. (1985) 'Real time ultrasonic scanning in the diagnosis of pregnancy and the estimation of gestational age in cattle', Veterinary Record 117, 5-8.

ULTRASONIC CHARACTERISTICS OF PATHOLOGICAL CONDITIONS OF THE BOVINE UTERUS AND OVARIES.

KÄHN, W. AND W. LEIDL
Gynaecological and Ambulatoric Animal Clinic, University of Munich
Königinstraße 12
D-8000 München 22
Germany

Introduction

Transrectal sonography has been widely used in the examination of physiological conditions of the female genital tract of the bovine for several years (Chaffaux et al. 1982, Tainturier et al. 1983, Pierson and Ginther 1984 a and b, Reeves et al. 1984, Kähn 1985, Taverne et al. 1985, White et al. 1985). The technology permits a safe, noninvasive pregnancy diagnosis as of week 4 and visualization of the embryo and foetus. Ultrasound scanning of the ovaries facilitates the identification of follicles and corpora lutea. These advantages were very promising for the use of ultrasound technology by practitioners as well as for research purposes.

More recently, sonography has been shown to be equally efficient in the evaluation of pathological conditions. This chapter deals with echogenic characteristics of clinically relevant uterine and ovarian disorders in the bovine.

Pathological findings on the uterus

EMBRYONIC LOSS

First indications for an imminent embryonic mortality are the small-for-date size of the embryo and decreased fluid of the conceptus. Conclusive evidence for the death of the embryo is provided by the cessation of heart activity.

In a number of pregnancies examined around day 15 to 40, retarded growth was observed for a few days while heart activity was still detectable (Fig. 1). Embryonic heart beat sometimes continued for a few days after the first indications of disturbed gestation such as retarded embryonic growth had been discovered (Fig. 2). It has been observed that heartrate was reduced immediately prior to the cessation of heart beat (Kastelic et al. 1988). An intact embryo at this stage of gestation has a heart rate of about 150 beats per minute or more. With progressing resorption the amount of placental fluids gradually decreased and their echogenity increased. The flaky reflections observed in the initial stages of embryonic loss intensified to give a snow-storm effect. The contours of the embryo became obscure.

In the past the occurrence of embryonic loss could be judged intra vitam only by indirect means. Ultrasound technology provides a safe method for the direct visualization of the conceptus, offering new possibilities in the study of embryonic death (Chaffaux et al. 1986).

MUMMIFICATION

In 2 cases of mummification the images of the uterine content were of limited diagnostic value. Highly echogenic structures were found directly adjacent to the uterine wall (Fig. 3). Less echogenic areas, as displayed by the foetal fluids suspending an intact foetus,

M. M. Taverne and A. H. Willemse (eds.), Diagnostic Ultrasound and Animal Reproduction, 53–65.

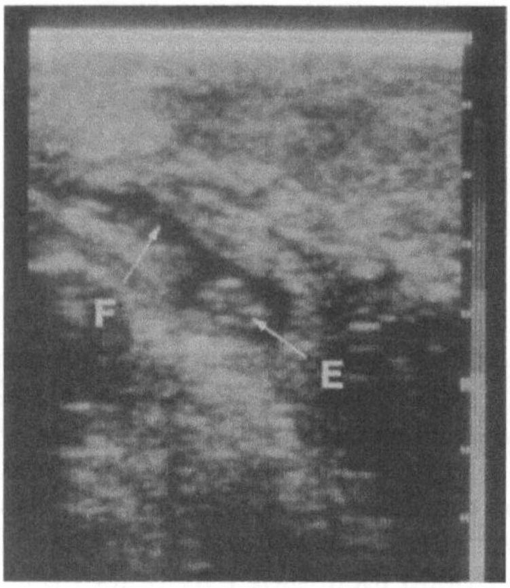

Fig. 1: Disturbed pregnancy 38 days after AI. Indications for impending embryonic death are reduced embryonic fluids (F) and the small-for-date size of the embryo (E).

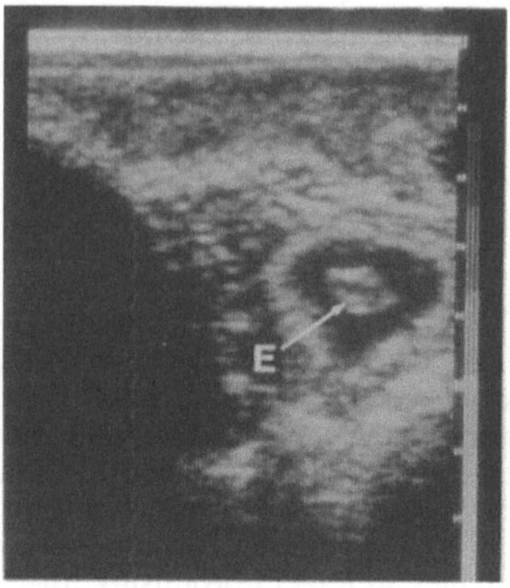

Fig. 2: Image of the embryo (E) depicted in Fig. 1 at day 43 of gestation after the death of the embryo. Embryonic heart activity had ceased and the conceptus was resorbed 10 days later.

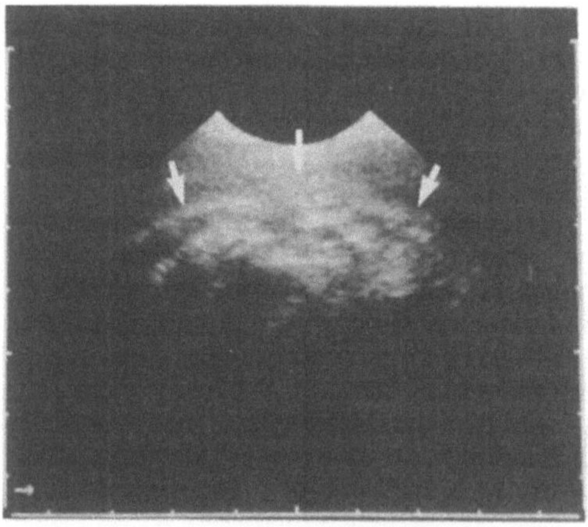

Fig. 3: Sonographic image of foetal mummification in a cow 348 days after AI. The transducer-near surface (arrows) of the mummified foetus is highly echogenic. Due to the strong absorption of sound waves, these are prevented from penetrating into deeper regions of the foetus.

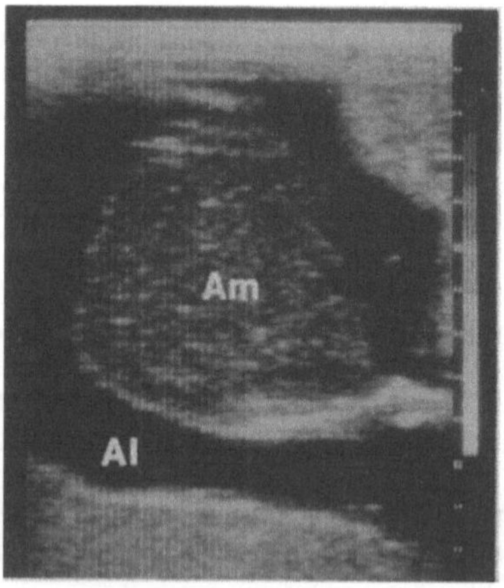

Fig. 4: Foetal maceration in a cow. The allantoic fluid (Al) is without echoes while the amniotic fluid (Am) shows reflections.

were not observable between the mummified foetus and the endometrium. No foetal organs or parts of the foetal body were identifiable during any of the investigations. The ultrasound waves penetrated only a short distance into the mummified foetus giving merely a narrow highly echogenic border. The deeper tissues reflected no ultrasound waves and therefore gave a near-black sonographic impression (Fissore et al. 1986). The process of mummification had apparently caused changes which led to an extensive absorption of the ultrasound waves in the superficial layers of the foetal remnants.

MACERATION

In addition to clinical symptoms ultrasonography may provide further information in cases of macerations, as seen in 3 cows. There was a very distinct difference in the echogenity of the allantoic and amniotic fluids (Fig. 4). While the allantoic fluid contained no reflecting particles and gave an almost black image, echogenic spots were observed whirling around in the amniotic fluid. The latter were probably a result of increased concentrations of cellular and tissue particles derived from the decomposing foetus.

Some foetal parts could be discerned within the echogenic amniotic fluid. Being macerated, the foetal outlines appeared very much less distinct than those of an intact living foetus. This was partly due to the high echogenity of the surrounding amniotic fluid which gave little contrast to the foetus, obscuring its contours. As a result of post mortal changes, the internal anatomy of the foetus was also less distinct. Only the bony structures, being of high tissue density and therefore extremely echogenic were still prominent (Fissore et al. 1986).

POSTPARTUM UTERUS

The most conspicuous structures detectable by ultrasound in the postpartum uterus are the caruncles (Fig. 5, 6 and 7). Various images of the caruncles can be obtained depending on the angle at which the ultrasound waves penetrate them. The superficial layers are echointensive while the main part of the caruncles is only slightly echogenic, typical for loosely packed soft tissue. Occasionally echogenic lines can be seen passing from the point of attachment on the uterine wall to the interior of the caruncle.

On the day following parturition, the uterine lumen often appears very small, apparently containing no large accumulations of fluids. In many cases, however, lochia are discovered in the uterine cavity (Fig. 6). The lochia display an echogenic pattern representative of fluids containing cellular particles. The caruncles then protrude like mushrooms into the relatively dark fluid-filled uterine lumen. Studies provide evidence that it is not exceptional to find a dilation of the uterus of a few centimeters even two weeks post partum (Okano and Tomizuka 1987). At this stage the caruncles had regressed normally, as can be demonstrated by ultrasound (Fig. 7). Small pockets of fluid are occasionally found even during the final stages of uterine involution up to week 4 post partum.

Under pathological postpartum conditions in the form of a lochiometra the uterus is often extremely dilated (Fig. 8) and the caruncles frequently not discernable. There is sometimes a hyperreflective zone towards the ventral aspect of the uterus. This is caused by the presence of tissue particles and cells which are most concentrated ventrally.

Some of the characteristics of the puerperal uterus can also be found under pathological conditions of the uterus such as endometritis. Thus the echogenity of lochial secretions may closely resemble that created by fluid accumulations observed in cases of endometritis (Fig. 9-11). The distinguishing criterion of the puerperal uterus is the presence of caruncles usually detectable throughout the early postpartum period. Other important aspects are the degree of dilation of the uterus and the incongruency in size between the previously

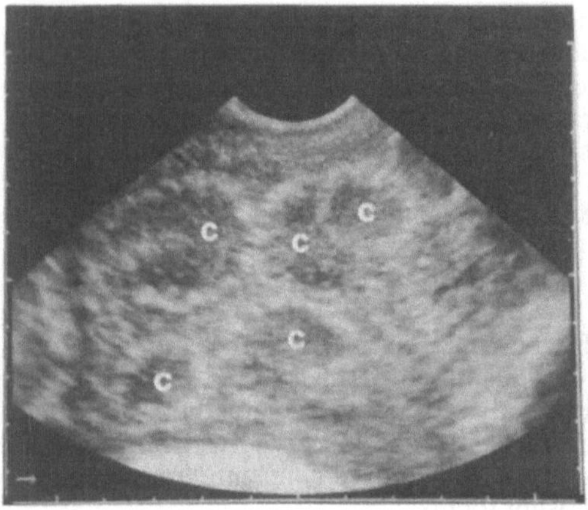

Fig. 5: Uterus with numerous caruncles (C) during normal uterine involution 2 days post partum. Note the conspicuous highly echogenic marginal region of the caruncles.

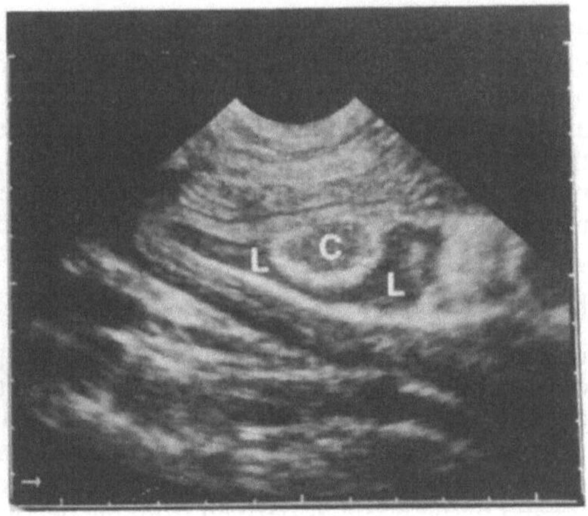

Fig. 6: Uterus with caruncle and some poorly echogenic lochia (L) in the normal postpartum uterus 10 days after parturition. The surface of the endometrium and the contours of one caruncle (C) are highly echogenic.

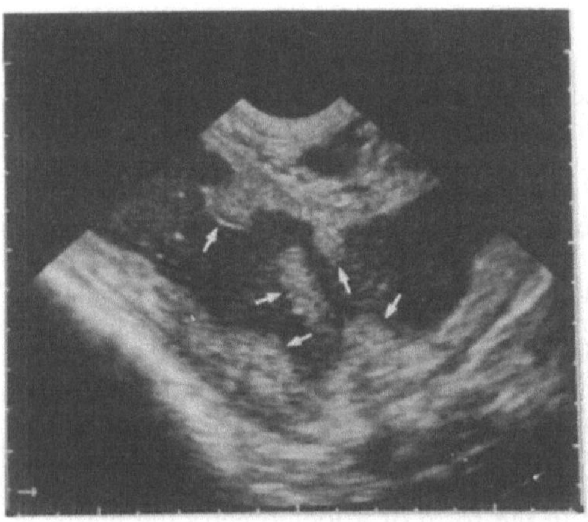

Fig. 7: Regressed caruncles (arrows) and large amounts of lochia 11 days post partum within the uterus of a cow with undisturbed puerperal period.

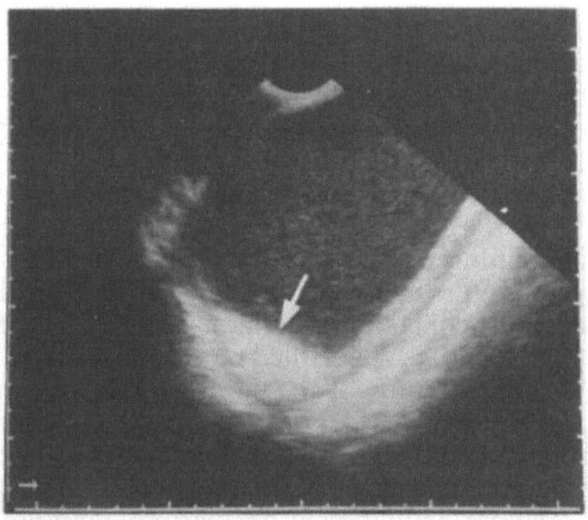

Fig. 8: Lochiometra of a cow with puerperal sepsis 12 days post partum. Due to the high cell content of the lochia, a hyperreflective zone is produced on the floor of the uterine lumen (arrow).

pregnant and non-pregnant uterine horns.

ENDOMETRITIS

The most apparent sonographic indication of endometritis is the unexpected accumulation of fluids in the uterine lumen. The amount of secretions varies greatly depending on the severity of the condition. In mild cases no uterine lumen, or merely a few fluid-filled pockets will be found on scanning the uterus (Fig. 9). Severe endometritis is characterized by the presence of frank collections of fluids which distend the entire uterus including both horns (Fig. 10).

The high echogenity of the fluids resulting from endometritis helps distinguishing them from physiologically occurring fluid collections such as oestrus secretions or the embryonic fluid of the early conceptus which are much less echogenic (Fissore et al. 1986). The echogenic patterns displayed by the products of inflammation range from small bright spots in mild cases to a "snow-storm effect" or, in severe cases, to almost bright white images on the screen. Real-time-scanning sometimes revealed turbulences within the larger collections of fluid.

Based on the experience acquired so far it appears that the usefulness of transrectal ultrasound technology for the diagnosis of endometritis depends on the degree of pathological changes. When only small collections of fluid are present, diagnosis is often not possible. On the other hand, severe cases causing the entire uterus to become distended are easily recognized.

An interesting phenomenon was observed after applying irritating solutions in cows with endometritis. Following the intrauterine infusion of a 4% iodine solution (Merckojod, Bayer Leverkusen), a sudden marked increase in echogenity of the endometrial surface was observed (Fig. 11). This echogenity was produced directly following the infusion and persisted for a few days.

The procedure was repeated on exenterated uteri. The results suggest that the increased echogenity was not caused by the iodine solution itself but rather by changes that had taken place within the endometrium. The degree of echogenity remained unaltered after the iodine solution had been removed from the lumen of the uteri.

PYOMETRA

Pyometra is characterized by large quantities of fluid often greatly dilating the uterus (Fig. 12). Diffuse flaky reflections of particles suspended in the fluid were seen on the ultrasonic screen. The degree of echogenity depended on the consistency of the pyometral fluid. When the uterine content was very thick and full of leucocytes and tissue debris, the echogenity closely resembled that of the uterine wall.

According to ultrasonographic findings, the thickness of the uterine wall can vary considerably. Thus thick-walled and thin-walled pyometras were found. Differentiation from a pregnancy on the basis of the thickness of the uterine wall is therefore only relevant if the affected uterine wall is extremely thick. Conclusive evidence for a pyometra is given, if the uterine content shows scattered echogenity of the fluid and the entire lumen of the uterus can be scanned without coming across any signs of an embryo or foetus.

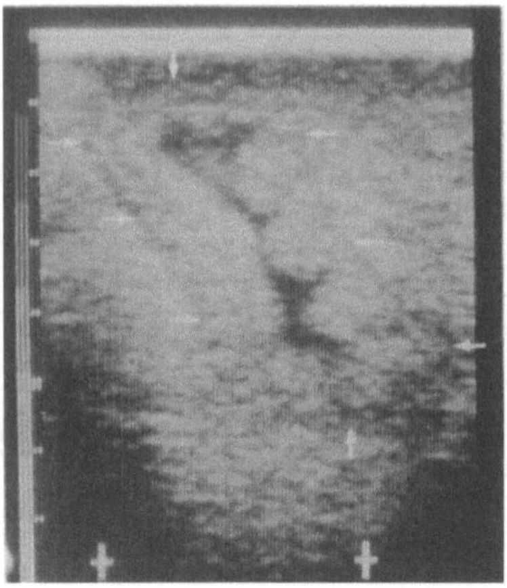

Fig. 9: Uterus with poorly echogenic secretions in the lumen as a result of chronic endometritis. The outline of the uterine horn is demarcated by arrows.

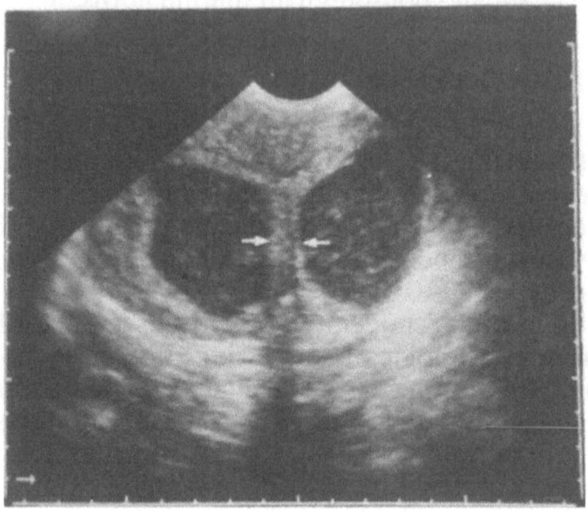

Fig. 10: Transverse section through a uterus containing a large volume of fluids as a result of chronic endometritis. The medial walls of the uterine horns meet in the middle (arrows). The lumen of each horn is occupied by fluid showing flaky reflections.

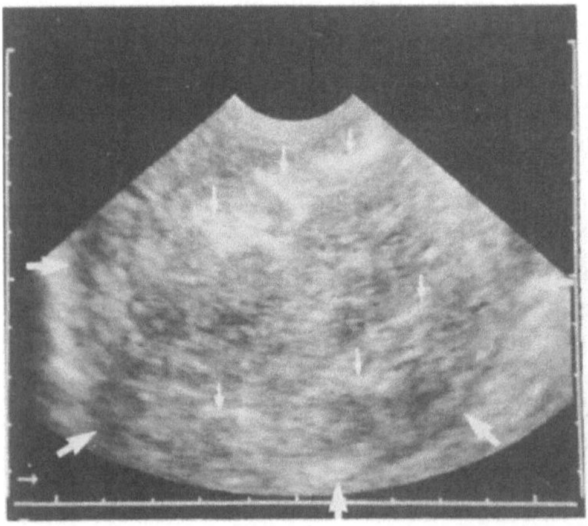

Fig. 11: Uterus showing intense echoes (small arrows) on the endometrial surface after infusion of an iodine solution (4% Merckojod solution). The large curvature is demarcated by large arrows.

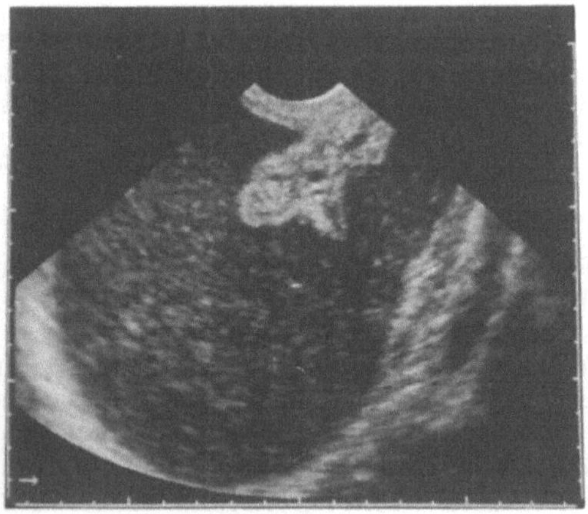

Fig. 12: Sagittal section through a pyometra. The uterus is greatly dilated and protrudes into the abdominal region (left side). Echo patterns of the secretions show the typical snow-storm effect.

Pathological findings on the ovaries

CYSTIC OVARIES

The ultrasonic anatomy of ovarian cysts and follicles may be very similar. Small structures with thin walls give no indication whether they represent cysts or follicles. The two most obvious differences are that cysts are larger and in cases of luteal cysts have thick walls.

Ovarian cysts generally contain large amounts of fluid which is poorly echogenic and therefore sonographically dark (Fig. 13-17). While the cavity of follicle-theca cysts is almost without reflections therefore appearing near black on the screen, luteal cysts occasionally show an internal network of echoes (Fig. 15).

Various shapes of ovarian cysts may be encountered including spherical, oval, polygonal as well as angular forms. Singular cysts are most frequently spherical. The shapes of adjacent cysts depends on their relative pressures (Fig. 14). In many cases pressures are equal and the apposed walls of adjacent cysts are more or less straight. When neighbouring cysts exert different pressures, those with higher pressures bulge into cysts with lower pressure. Therefore small vesicles usually form indentations in larger adjacent cysts.

Sonographically detectable ovarian cysts can be divided into 2 categories according to the morphology of their wall. Cysts with very thin walls showing no particular organization are assumed to be follicle-theca cysts (Fig. 13).

Cysts having walls a few mm in thickness with the echogenity of luteal tissue are categorized as follicle-luteal cysts (Fig. 15). The ovarian stroma surrounding these cysts is more echogenic than the cyst wall. When the features concerning the walls of the two types of cysts are characteristically developed, i. e. when follicle-theca cysts have very thin walls and follicle-luteal cysts have thick luteinized walls, sonographic differentiation presents no problem. However, it is known that transitional forms occur, and that about 34% of so-called follicle-theca cysts show some degree of luteinization within their walls (Leidl et al. 1979). Sonographic identification of these intermediate forms is therefore not always possible.

The presence of cystic structures on bovine ovaries does not imply a pathological condition in all cases. Using ultrasound technology the concurrent presence of ovarian cysts and a cyclic corpus luteum could be objectified immediately (Fig. 16). Furthermore, it is well known that follicular cysts may be present during normal pregnancy (Fig. 17 and 18).

Generally, ovarian cysts are identifiable sonographically by their large, nonechogenic interior created by the cystic fluid (Kähn and Leidl 1986). The diagnosis of follicle cysts with a thin, echogenic wall is usually not difficult. Follicle-luteal cysts are also readily recognized if the diameter is more than 40 mm and the luteinized wall is thinner than 5 mm. However, small follicle-luteal cysts with relatively thick walls could be mistaken for large corpora lutea with cavities. Ultrasound examination may also be inconclusive in the rare cases when large fluid-filled cavities of corpora lutea surrounded by a relatively thick wall are found. Criteria for the differentiation are presented in the following section.

CYSTIC CORPORA LUTEA

Corpora lutea with signs of cavitation are known as cystic corpora lutea. Up to now no convincing evidence for a pathological significance of these structures has been found. In fact, a number of recent findings support the assumption, that these structures present a normal variation of corpora lutea in the bovine (Kähn 1986, Kito et al. 1986, Kähn and Leidl 1988). These studies have shown no negative influence on hormone profiles or pregnancy rate of affected animals. Hence, it would not be correct to declare cystic corpora lutea as ovarian cysts and a clear differentiation between these two structures becomes important.

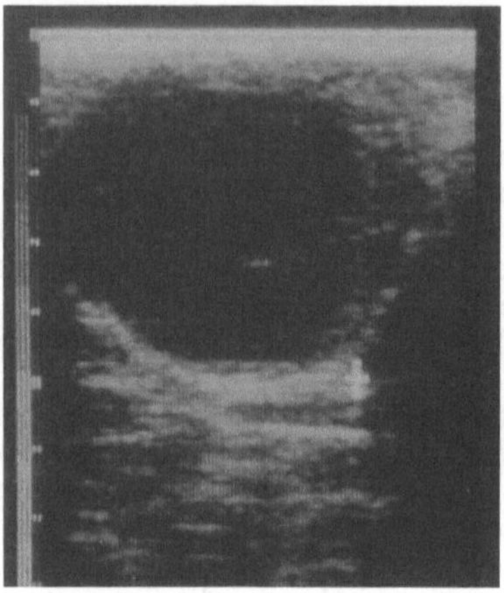

Fig. 13: Bovine ovarian follicular cyst with a diameter of 5 cm.

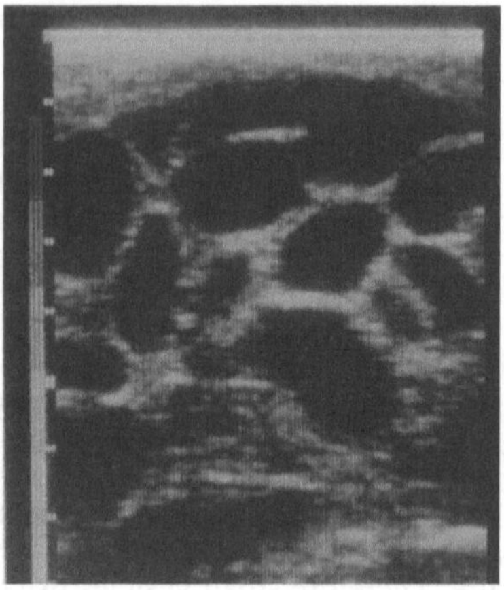

Fig. 14: Ovary of a cow which developed cystic degeneration of the ovaries 7 days after the last FSH-injection to induce superovulation.

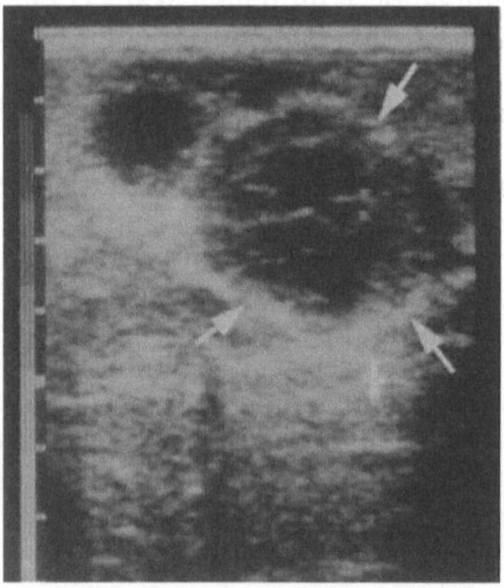

Fig. 15: Ovary with a follicle-luteal cyst (arrows). The luteinized wall is 3-4 mm thick. Note the network of echoes within the cavity of the cyst.

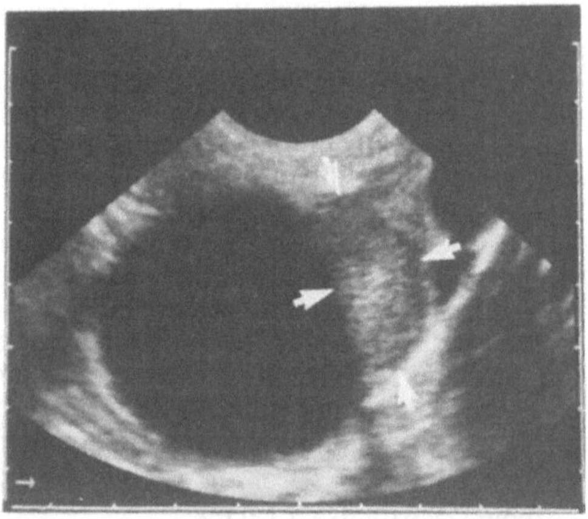

Fig. 16: Ovary with a follicular cyst and a cyclic corpus luteum (arrows) in a cow that had been treated for ovarian cysts 13 days previously with GnRH (20 µg Buserelin i.m.). Progesterone was high on the day of sonography. PGF-2_a was injected and A.I. 4 days later resulted in a pregnancy.

Corpora lutea with a cavity are comprised of a thick wall of low echogenity surrounding a central, fluid-filled zone (Fig. 19). The luteal tissue of the wall produces the same echogenity as luteal tissue of compact corpora lutea.

The central area is usually oval in shape, seldom round, and is located in the middle of the corpus luteum. The largest diameters of the fluid-filled core may vary from a few mm to about 1.5 cm. Seldom do they reach dimensions of 2.0 cm or more. The echogenity of the central cavity, being fluid-filled, is similar to that of follicles. Slight reflections could be found in a few cases (Fig. 20). The wall of the cystic corpora lutea serves as a criterion of differentiation from ovarian follicles. Whilst the fluid-filled antrum of follicles is separated from the highly echogenic ovarian tissue by only a thin wall, the fluid-filled cavities of cystic corpora lutea are surrounded by a thick wall of luteal tissue which is of weak echogenity.

So called cystic corpora lutea have been found during normal oestrus cycles and in the first weeks of pregnancy. During the first 10 days following ovulation, cows with normal cycles and cows that had conceived were examined for the presence of cystic corpora lutea. Half of these cows, irrespective of whether they had become pregnant or not, showed corpora lutea with cavities >3 mm; one third of them had cavities >7 mm. The cavities usually decreased after day 10 (day 0 = day of ovulation) post ovulation and had frequently disappeared by week 3. After this stage, it is exceptional to find cystic corpora lutea during early pregnancy.

For a clear differentiation between follicle-luteal cysts and cystic corpora lutea, various criteria must be considered. These are overall size, shape of the vesicle and the cavity, thickness of the luteinized wall and echogenity of the interior area. Cystic corpora lutea are usually no larger than 3 cm in diameter and the wall is about 5-10 mm thick. Corpora lutea are rarely spherical, usually presenting an oval shape on the screen. The image of the cavities depend on the direction of the ultrasound beam and is either circular or oval. The fluid-filled cavities of corpora lutea only seldom present reflections, more often being homogeneous and near-black, whereas reflections are frequently found in follicle-luteal cysts. In ambiguous cases, repeat examination after one week should bring clarity. A cystic corpus luteum will show cyclic development such as thickening of the luteal wall and diminution of the cavity.

OVARIAN TUMOURS

Ovarian tumours occur very rarely in cattle. This is a case report of an ovarian tumour in a cow examined by ultrasound.

The cow was 3 1/2 years old, had calved 2 months previously and showed no clinical signs of unsoundness. On rectal examination prior to artificial insemination, a compact structure, the size of 2 children's heads was palpated. During the following weeks the cow showed nymphomaniac behaviour and her plasma progesterone level was consistently <1.0 ng/ml.

The sonographic image of the tumour in the left ovary was characterized by two distinct zones (Fig. 21). Dorsally, the nonechogenic sections of numerous large blood vessels were depicted. The remainder of the tumour showed a coarse-grained echogenic pattern typical for a solid mixed cell type neoplasm and was penetrated by many very small blood vessels. The mingling of bright and moderate echoes reflects the firm consistency of the tumour in which circumscribed areas are embedded. The contours of the tumour were recognizable and its dimensions could be registered.

Pathological examination of the tumour after its removal by ovariectomy revealed a granulosa-cell tumour on the left ovary, the size of a football, supplied by a few blood vessels. The right ovary also, was affected by neoplastic changes resembling a granulosa-

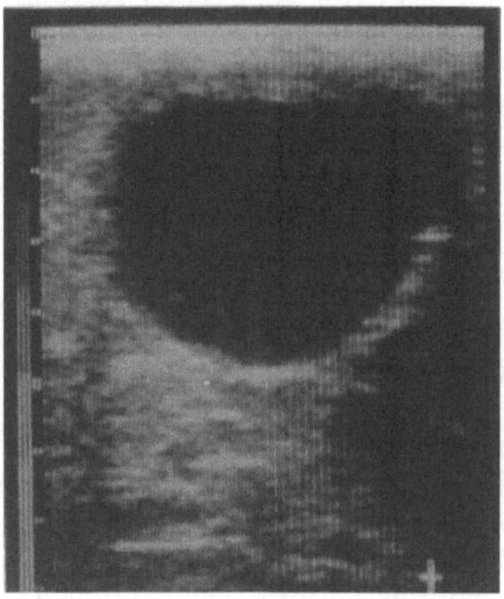

Fig. 17: Ovarian cyst of a cow on day 59 of pregnancy (corresponding pregnant uterus see Fig. 18).

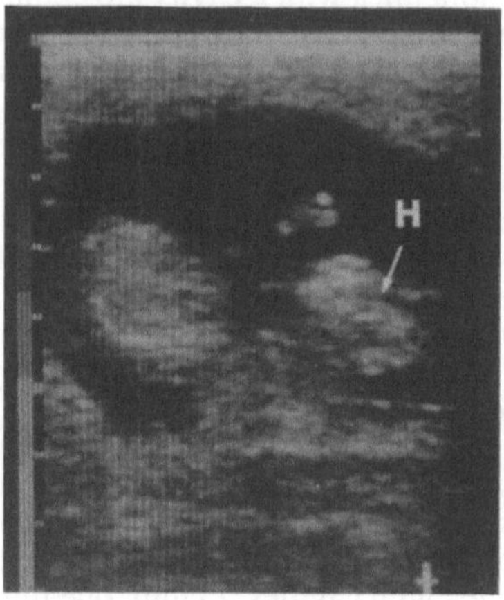

Fig. 18: Pregnant uterus and foetus in a cow with an ovarian cyst (depicted in Fig. 17) on day 59 after ovulation. The head (H) of the foetus is positioned to the right, the forelimbs point upwards and the trunk is on the left.

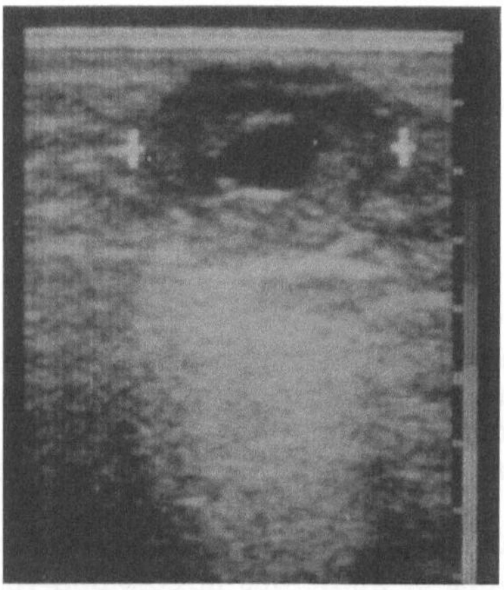

Fig. 19: Cystic corpus luteum (between the 2 white crosses) 8 days after ovulation. The corpus luteum measures 36 x 27 mm, the cavity 13 x 9 mm.

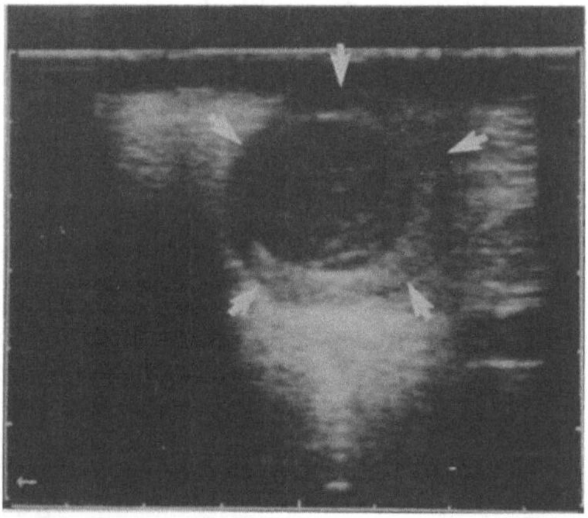

Fig. 20: Cystic corpus luteum (arrows) with a large cavity (22 x 20 mm) 14 days post ovulation. The succeeding oestrus began on the expected date. The fluid-filled cavity of corpus luteum shows slight reflections.

cell tumour.

This description of a granulosa-cell tumour is not necessarily typical for this type of ovarian tumour in the cow. In this case, the tumour was very compact, the highly echogenic cross-sectional image being interrupted only by nonechogenic blood vessels. Apparently, granulosa-cell tumours containing numerous fluid-filled cavities resembling follicles also occur in the bovine ovary (Andresen et al. 1986). The ultrasound image of this type of tumour would presumably resemble more closely the appearance of granulosa-cell tumours found in mares (Kähn and Leidl 1987). These are frequently multi-cystic and encapsulated by connective tissue (White and Allen 1985). The content of these cysts is either serous or haemorrhagic giving a near-black or moderately echogenic ultrasound image.

Conclusions

Ultrasound technology facilitates the visualization of various conditions of the uterus involving dilatation of the uterine lumen by fluids. Nonechogenic fluids occur only under physiological conditions such as oestrus secretions or the placental fluids of embryonic vesicles (Pierson and Ginther 1987). Fluids which accumulate as a result of pathological changes always show some echogenity. Echogenic patterns range from disseminated bright echoes to the snow-storm effect producing an almost white image on the screen. The degree of ultrasound reflections depends on the concentration of particles suspended in the secretions. Increased echogenity of intra-uterine fluid collections is always present in chronic endometritis, pyometra, maceration and in the postpartum uterus. After parturition it is observed in pathological as well as physiological uterine involution.

The availability of ultrasound technology permits an improved examination of the bovine ovary. Measurement of the size of vesicles, registration of the echogenity of the wall and the antrum can often provide the information for an exact diagnosis of follicles, ovarian cysts and corpora lutea with and without cavities. Also the influence of hormone therapy in the case of ovarian disorders can be observed.

In conclusion it can be said that transrectal ultrasound scanning is a valuable method of diagnosing pathological conditions of the uterus and ovaries in the bovine. In some cases it gives supplementary evidence when classical methods such as rectal palpation, vaginal examination or hormone analysis cannot provide sufficient information.

References

Andresen, P., Dehning, R. und Scheidegger, A. (1986) 'Ein Beitrag zum Granulosazelltumor beim Rind', Prakt. Tierarzt 4, 307-309.

Chaffaux, S., Valon, F. et Martinez, J. (1982) 'Evolution de l'image échographique du produit de conception chez la vache', Bull. Acad. vét. Fr. 55, 213-221.

Chaffaux, S., Reddy, G. N. S., Valon, F. and Thibier, M. (1986) 'Transrectal real-time ultrasound scanning for diagnosing pregnancy and for monitoring embryonic mortality in dairy cattle', Anim. Reprod. Sci. 10, 193-200.

Fissore, R. A., Edmondson, A. J., Pashen, R. L., and Bondurant, R. H. (1986) 'The use of ultrasonography for the study of the bovine reproductive tract. II. Non-pregnant, pregnant and pathological conditions of the uterus', Anim. Reprod. Sci. 12, 167-177.

Kähn, W. (1985) 'Zur Trächtigkeitsdiagnose beim Rind mittels Ultraschall', Tierärztl. Umsch. 40, 472-477.

Kähn, W. (1986) 'Vorkommen und Wachstum von Gelbkörpern mit Hohlraum während des Ovarialzyklus bei Rindern und deren Hormonprofile', Dtsch. tierärztl. Wschr. 93,

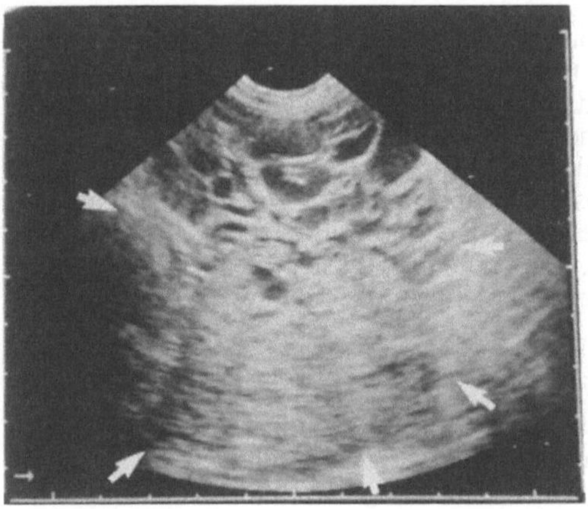

Fig. 21: Ovary with a granulosa-cell tumour (arrows). Nonechogenic sections of numerous large blood vessels are visible in the upper region of the tumour. The ventral part of the tumour shows an echogenic pattern typical for structures of relatively high tissue density.

475-480.
Kähn, W. und Leidl, W. (1986) 'Die Anwendung der Echographie zur Diagnose der Ovarfunktion beim Rind', Tierärztl. Umsch. 41, 3-12.
Kähn, W. und Leidl, W. (1987) 'Echographische Befunde an Ovarien von Stuten', Tierärztl. Umsch. 42, 257-266.
Kähn, W. und Leidl, W. (1988) 'Auswirkungen einer HCG-Stimulation bei Rindern mit einem zystischen oder einem kompakten Gelbkörper: Progesteronprofile, Ovulationsinduktion und Ausbildung eines 2. Gelbkörpers', Tierärztl. Umsch. 43, 430-439.
Kastelic, J. P., Curran, S., Pierson, R. A. and Ginther, O. J. (1988) 'Ultrasonic evaluation of the bovine conceptus', Theriogenology 29, 39-54.
Kito, S., Okuda, K., Miyazawa, K. and Sato, K. (1986) 'Study on the appearance of the cavity in the corpus luteum of cows by using ultrasonic scanning', Theriogenology 25, 325-333.
Leidl, W., Braun, U., Stolla, R. und Bostedt, H. (1979): Zur Ovarialzyste des Rindes. I. Klassifizierung und Diagnose. Berl. Münch. tierärztl. Wschr. 92, 369-376.
Okano, A. and Tomizuka, T. (1987) 'Ultrasonic observation of postpartum uterine involution in the cow. Theriogenology 27, 369-376.
Pierson, R. A. and Ginther, O. J. (1984 a) 'Ultrasonography for detection of pregnancy and study of embryonic development in heifers', Theriogenology 22, 225-233.
Pierson, R. A. and Ginther, O. J. (1984 b) 'Ultrasonography of the bovine ovary. Theriogenology 21, 495-504.
Pierson, R. A. and Ginther, O. J. (1987) 'Ultrasonographic appearance of the bovine uterus during the estrous cycle', J. Am. vet. med. Ass. 8, 995-1001.
Reeves, J. J., Rantanen, N. W. and Hauser, M. (1984) 'Transrectal real-time ultrasound scanning of the cow reproductive tract', Theriogenology 21, 485-494.
Tainturier, D., Andre, F., Chaari, M., Sardjana, K. W., Le Net, J. L. et Lijour, L. (1983) 'Intérêt de l'échotomographie pour le contrôle de la reproduction d'un grand troupeau de vaches laitières' Rev. Méd. vét. 134, 419-424.
Taverne, M. A. M., Szenci, O., Szétag, J. and Piros, A. (1985) 'Pregnancy diagnosis in cows with linear-array real-time ultrasound scanning: a preliminary note', The Veterinary Quarterly 7, 264-270.
White, R. A. S. and Allen, W. R. (1985) 'Use of ultrasound echography for the differential diagnosis of a granulosa cell tumour in a mare' Equine vet. J. 17, 401-402.
White, I. R., Russel, A. J. F., Wright, I. A. and Whyte, T. K. (1985) 'Real-time ultrasonic scanning in the diagnosis of pregnancy and the estimation of gestational age in cattle', Vet. Rec. 117, 5-8.

EARLY PREGNANCY DIAGNOSIS IN CATTLE BY MEANS OF TRANSRECTAL REAL-TIME ULTRASOUND SCANNING OF THE UTERUS.

A.H. Willemse and M.A.M. Taverne
Department of Herd Health and Reprodution
State University of Utrecht
Yalelaan 7
3584 CL Utrecht, The Netherlands

INTRODUCTION

Since the number of days between calving and subsequent conception determines the calving interval in cattle, early pregnancy diagnosis has become a key to economic succes. Pregnancy is generally diagnosed by palpating the uterus and its contents per rectum. The accuracy of this method and the earliest time of which pregnancy can be recognized is dependent upon the experience of the palpator and on the pregnancy diagnosis criteria used.

Use of transrectal ultrasonography of the uterus for diagnosis of pregnancy has been reported by several authors during recent years. Using a 3.5 MHz transducer, Chaffaux (1982) observed irregular, non-echogenic structures in the uterine lumen from day 28 after insemination. From 35 days on, the embryo could be located in this vesicle. Modern and more powerful equipment (7.5 MHz) enabled Boyd et al., (1988) to detect pregnancy as early as 9 days post insemination by imaging a vesicle within the lumen of the uterine horn. The embryo was seen within this vesicle at day 13 and the earliest heart beat was detected in the embryo at day 22.

Besides studies on early pregnancy detection, this technique can be used to study the ultrasonic appearance of the conceptus (Curran et al., 1986 a, b; Kastelic et al., 1988), for estimation of the gestational length in beef cattle (White et al., 1985), for the determination of foetal sex (Müller and Wittkowski., 1986), for the estimation of embryonic loss (Chaffeux et al., 1986; Wilson, 1988) and to study uterine morphology in normal and pathological conditions (Fissore et al., 1986). Besides these descriptive studies only a limited number of reports are available on the reliability and accuracy of this technique for early pregnancy diagnosis (Taverne et al., 1985, Chaffeux et al., 1986 and Hanzen and Delsaux, 1987).

The present study was undertaken with the objective of determining the predictive accuracy of transrectal ultrasound scanning of the uterus for early pregnancy diagnosis under field conditions.

M. M. Taverne and A. H. Willemse (eds.), Diagnostic Ultrasound and Animal Reproduction, 67–72.

MATERIALS AND METHODS

Two to ten year old cows at 3 different state farms in Hungary were used in this experiment. The cows were selected in such a way that during a period of 4 days a maximum number of animals, probably in early stages of pregnancy, could be examined. Cows (n=85) not seen in heat again after insemination were tested between 26 and 33 days after A.I.

A portable scanner type 400 (Pie Medical, Maastricht, the Netherlands) with a 5 MHz rectal linear-array transducer was used, reaching a depth of 10-12 cm from the surface of the probe. After removal of the faeces the transducer was inserted into the rectum.

The transducer was held with its longitudinal axis parallel to that of the cow and excursions from the midline to both the left and the right sides were made to image the entire uterus. After each examination the transducer was cleaned.

A cow was considered to be pregnant when on the monitor an irregularly shaped, non-echogenic black shadow was seen within the uterine lumen, representing the fluid filled allantoic cavity. Visualization the embryo itself was not attempted. Pregnancy was confirmed by rectal palpation between 50 and 60 days after A.I.; calving data were recorded.

Knowing the actual calving dates the results were arranged as follows: correct positive diagnoses (a), incorrect positive diagnoses (b), correct negative diagnoses (c), and incorrect negative diagnoses (d). From these, the sensitivity (100 x a/(a+d), the specificity (100 x c/c+b), the positive predictive value (100 x a/(a+b) and the negative predictive value (100 x c/(c+d) of the test were calculated.

RESULTS

A well involuted non-pregnant uterus can be easily visualized on the screen (fig. 1). The shape of the coiled uterine horns can be recognized and no apparent lumen is present. Fig.2 shows the echographic picture of an early pregnant uterus, 26 days after A.I. The irregular shaped non-echogenic area visible in the uterine lumen represents the allantoic fluid.

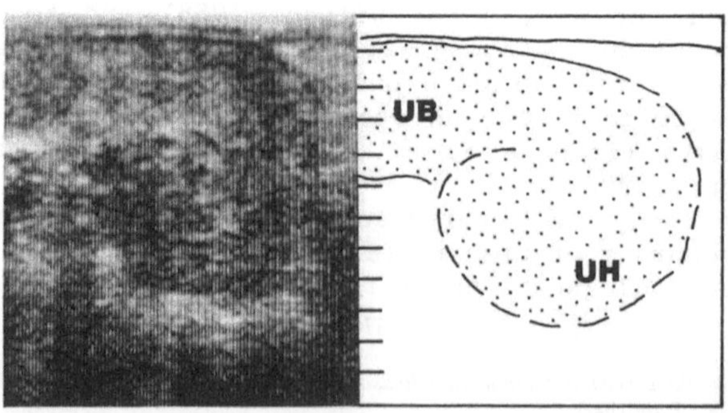

Figure 1. Echographic image of a non-pregnant uterus (UB: uterine body; UH: uterine born).

DATA FROM

	Present Study		Hansen & Delsaux (1987)			Chauffeux x.s. (1986)			Taverne c.s. (1985)	
	days after A.I.		days after A.I.			days after A.I.			days after A.I.	
	26–29 n=48	30–33 n=37	< 30 n=21	30–39 n=126	40–49 n=93	< 30 n=68	30–39 n=69	40–49 n=64	21–35 n=58	36–49 n=63
diagn. pregnant correct (a)	25	18	12	91	73	13	32	32	27	43
diagn. pregnant incorrect (b)	3	2	5	7	2	16	10	10	0	3
diagn. not pregnant correct (c)	20	16	3	26	17	32	25	22	24	17
diagn. not pregnant incorrect (d)	0	1	1	2	1	8	1	0	7	0
sensitivity a/a+d x 100	100	95	92	97	98	62	97	100	79	100
specificity c/c+b x 100	87	89	37	78	89	67	71	69	100	85
pos.predictive value a/a+b x 100	89	90	70	92	97	45	76	76	100	93
neg.predictive value c/c+d x 100	100	94	75	92	94	80	96	100	77	100

Table 1. A comparison between the accuracies of pregnancy tests performed by means of two-dimensional ultrasonography.

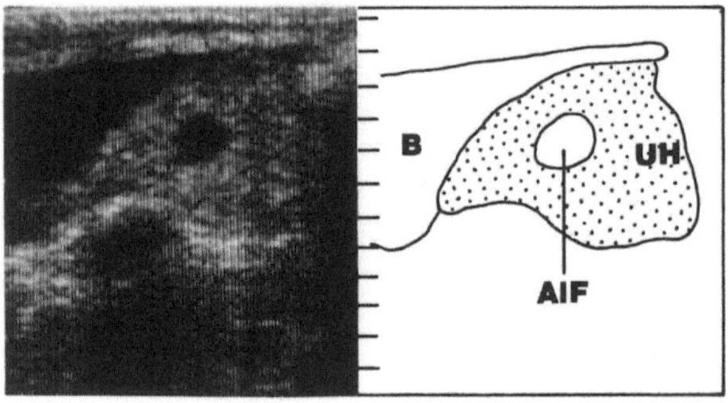

Figure 2. Echographic image of a pregnant uterus at 26 days after A.I. (B: bladder; ALF: allantoic fluid; UH: uterine born).

To compare the results of our test with those of previous reports (table 1) we devided the 85 animals into two groups: cows that were examined between days 26 and 29 after insemination (n=48) and cows examined between 30 and 33 days after A.I. (n=37).

Among the 25 animals in which pregnancy was confirmed later, each was diagnosed correctly by scanning the uterus before day 30. Among the 23 non calving animals, 20 (87%) were diagnosed correctly before day 30. With scanning between day 30 and day 33 95% (18 from the 19) of the pregnant cows were correctly diagnosed as pregnant while 89% (16 from the 18) of the non calving cows had been diagnosed as non pregnant.

DISCUSSION

The main purpose of pregnancy examinations is to detect open animals as early as possible. Based on the results of an early diagnosis one may consider oestrus induction and reinseminatation; sometimes a decision to cull will be made. The negative predictive value of an early diagnosis should therefore be 100% to avoid early abortion or culling of pregnant animals. Table 1 shows that we made no false negative diagnoses between days 26 and 29 after A.I. while in most of the other studies the negative predictive value approached 100% only after day 30. Taverne et al., (1985) used a 3.0 MHz probe. This difference may partly explain the rather high number (20%) of false negative diagnoses in our earlier report; 4 of the 7 false negative diagnoses in that study were made before day 30. We therefore conclude that the use of a 5 MHz transducer is necessary to adequately improve the sensitivity of this diagnostic technique. If the results of all studies with a 5 MHz transducer and which are listed in table 1 are taken together, only 5 of the 251 pregnant cows (2.0%) investigated after day 30 were incorrectly diagnosed.

The positive predictive value is considerably lower than the negative predictive value in all studies: after examination with a 5 MHz probe later than 30 days, 277 animals were considered to be pregnant; 31 of these cows (11,2%) appeared to be not in calf after rectal palpation later on. Mortality of the conceptus at some stage after the day of ultrasonographic examination may explain a great part of this difference. In a group of 100 inseminated cows Chaffaux et al., (1986) found 14 animals with an elevated level of progesterone in milk or plasma on day 23 after insemination and a positive ultrasound test on day 35, but they appeared to be not pregnant at rectal palpation on day 60. Assuming the animals were not inseminated in the luteal phase of the cycle, late embryonic mortality is undoubtedly responsible for this discrepancy.

It may be concluded from this study that intra rectal real-time ultrasound scanning of the uterus is an accurate method for pregnancy diagnosis after day 30. Although still based on a limited number of animals, the results from our scans made between days 26 and 30 indicate that even before day 30 the technique is very reliable.

REFERENCES

Boyd, J.S., Omran S.N. and Ayliffe T.R. (1988) "Use of a high frequency transducer with real time B-mode ultrasound scanning to identify early pregnancy in cows", Veterinary Record 123, 8-11.

Chaffeux, S., Valon F. and Martinez J. (1982) "Evoluation de l'image échograpique du produit de conception chez la vache", Bulletin Académie Véterinaire de France 55, 213-221.

Chaffeux, S., Reddy G.N.S., Valon F. and Thibier M. (1986) "Transrectal real-time ultrasound scanning for diagnosing pregnancy and monitoring embryonic mortality in dairy cattle" Animal Reproduction Science 10, 193-200.

Curran, S., Pierson R.A. and Ginther O.J. (1986) "Ultrasonic appearance of the bovine conceptus from days 10 through 20", Journal American Veterinary Medical Association 189, 1289-1294.

Curran, S., Pierson R.A. and Ginther O.J. (1986) "Ultrasonic appearance of the bovine conceptus from days 20 through 60", Journal American Veterinary Medical Association 189, 1295-1302.

Fissore, R.A., Edmondson A.J., R.L. Pashen and R.H. Bondurant 1986) "The use of ultrasonography for the study of the bovine reproductive tract. II. Non-pregnant, pregnant and pathological conditions of the uterus", Animal Reproduction Science 12, 167-177.

Hanszen, C. and Delsaux B. (1987) "Use of transrectal B-mode ultrasound imaging in bovine pregnancy diagnosis", Veterinary Record 121, 200-202.

Kastelic, J.P., Curran, S. Pierson R.A. and Ginther O.J. (1988) "Ultrasonic evaluation of the bovine conceptus", Theriogenology 29, 39-54.

Müller, E. and Wittkowski, G. (1986) "Visulization of male and female characteristics of bovine fetuses by real-time ultrasonics", Theriogenology 25, 571-574.

Taverne, M.A.M., Szenci O., Szétag J. and Piros A. (1985) "Preganancy diagnosis in cows with linear-array real-time ultrasound scanning: a prelominary note", The Veterinary Quarterly 7, 264-270.

White, I.R., Russel, A.J.F., Wright I.A. and Whyte T.K. (1985) "Real-time ultrasonic scanning in the diagnosis of pregnancy and the estimation of gestational age in cattle", Veterinary Record

Wilson, J.M. and Zalesky D.D. (1988) "Early pregnancy determination in the bovine utilizing ultrasonography", Theriogenology 29, 330. 117, 5-8.

THE APPLICATION OF REAL-TIME ULTRASONIC SCANNING IN COMMERCIAL SHEEP, GOATS AND CATTLE PRODUCTION ENTERPRISES

A. J. F. RUSSEL
Macaulay Land Use Research Institute,
Bush Estate, Penicuik,
Midlothian EH26 OPY,
Scotland, U.K.

ABSTRACT. Real-time ultrasonic scanning can be used to diagnose pregnancy and determine foetal numbers in sheep, goats and cattle, and has advantages over other approaches in terms of safety, accuracy and rapidity. In sheep and goats the principal benefits of the use of the technique include the ability to relate feed inputs in late pregnancy to foetal numbers, thus improving the efficiency of use of feedstuffs. Feeding in relation to foetal numbers also offers a means of manipulating birthweights and can lead to substantial reductions in mortality of both offspring and their dams. In cattle real-time scanning can also be used to estimate foetal age with a high degree of accuracy thus enabling cows to be managed more efficiently in relation to expected date of calving.

Introduction

Real-time ultrasonic scanning provides a safe, accurate and rapid means of diagnosing pregnancy, estimating foetal age and determining foetal numbers in sheep and other species of farm livestock. Although ultrasound has been used to a limited extent over the last ten years or more to diagnose pregnancy in horses, the impetus to apply the technology developed for human medicine to UK farm livestock came from the sheep industry. The very large differences in nutrient requirements between pregnant and non-pregnant ewes, and between ewes carrying different numbers of foetuses, had been appreciated for some time (Thomson and Aitken, 1959; Russel, Doney and Reid, 1967) but for many years the lack of a practicable technique for diagnosing pregnancy and determining foetal numbers had made it impossible to apply this knowledge in practice.

The inability to use existing knowledge on the nutrition and management of the pregnant ewe has been a source of significant economic loss to the UK sheep industry. The feeding of non-pregnant ewes as if they were in-lamb is clearly inefficient and wasteful. Without a knowledge of foetal numbers the feeding of pregnant ewes is a matter of guesswork or, at best, compromise, with detrimental effects to both production and economic efficiency. Lamb mortality is closely related to birthweight, with both underweight and overweight lambs being at risk. Twin or triplet lambs from ewes fed as if they were carrying only singles will be

M. M. Taverne and A. H. Willemse (eds.), Diagnostic Ultrasound and Animal Reproduction, 73–87.

underweight and likely to die because of weakness and excessive heat loss, while single lambs from ewes fed as if they were carrying multiples will be overweight and at risk as a result of a prolonged and difficult birth. In flocks where all ewes are treated alike those carrying more than the average number of lambs are also more likely to develop metabolic disorders, such as pregnancy toxaemia, in late pregnancy.

The task of the flockmaster or shepherd at lambing is also made difficult if ewes cannot be managed according to the number of foetuses carried. In extensive systems of management in particular, much time and effort can be spent in dealing with the problems of weak and underweight lambs, orphan lambs, ketotic ewes and ewes with little or no milk, all resulting from inadequate nutrition in late pregnancy.

Although similar considerations apply to goats, the demand for an accurate means of early pregnancy diagnosis in this species stems from recent developments in goat fibre production in the UK. Numbers of both cashmere and Angora goats have been increasing rapidly in recent years. Several thousand dairy goats are used annually as recipients for imported frozen embryos or for fresh embryos obtained in multiple ovulation and embryo transfer programmes. The efficient running of these programmes requires that non-pregnant recipients are identified as soon as possible so that they can be used again in the same breeding season. Also, when pregnant recipients carrying potentially valuable foetuses are sold, it is important to both the vendor and purchaser that the number of foetuses borne by each surrogate mother is accurately determined.

The UK beef cattle industry also has requirements for early pregnancy diagnosis and the estimation of foetal age, and in the future is likely to demand a means of determining foetal numbers. Most beef cows are mated by natural service and in many herds the bulls are run with the cows for several months. Fertility problems in either sex can be unrecognized for considerable periods of time, leading to serious economic loss. Early pregnancy diagnosis is important in the identification of such problems and in attempts to improve the efficiency of production by reducing both the calving interval and the duration of the calving season. The management of herds of beef cows can also be improved by grouping animals in relation to their expected date of calving (Russel and Wright, 1986). Recent developments in the in vitro fertilisation and culture of bovine embryos is likely to lead to a substantial proportion of cows carrying twins, and a means of determining foetal numbers will therefore become an essential tool in the nutritional management of herds of beef cows.

Non-ultrasonic Pregnancy Diagnosis

In theory it is possible to diagnose pregnancy and determine foetal numbers in sheep, goats and cattle by estimating the concentrations of certain hormones and metabolites in the blood. Although this approach is being used successfully in dairy cows where precise service or insemination dates are known, it is, for a variety of reasons, impracticable in large sheep flocks or herds of goats or beef cattle.

Pregnancy is normally diagnosed in cattle by a clinical examination in which the uterus is palpated through the rectal wall. Such examinations are normally made two months or more after the end of the mating period (Roberts, 1971). While rectal palpation is likely to remain a valuable and widely practised technique in cattle, the emphasis now being placed on improved reproductive efficiency demands an earlier means of accurate diagnosis. A rectal-abdominal palpation technique for diagnosing pregnancy in sheep has been described by Hulet (1973) but this has proved to be inaccurate (Trapp and Slyter, 1983) and entails the severe risk of rectal perforation (Turner and Hindson, 1975).

X-ray techniques, such as those used by Wenham and Robinson (1972) and reviewed by Wilson (1981), and the more sophisticated radiological technology described by Beach (1982), are capable of providing high levels of accuracy of both pregnancy diagnosis and determination of litter size in sheep. The capital cost of such equipment is high, it is generally mobile but not portable, and there is always a potential health hazard to operators working with perhaps tens of thousands of animals each year. For these reasons and because such techniques can be used successfully only after about 80 days of gestation, it is unlikely that they will find widespread application in commercial farming enterprises.

Non-ultrasonic techniques such as measurement of the viscosity of cervical mucus, arborization patterns of cervical-vaginal mucus, abdominal palpation, and others considered by Richardson (1972) as unsatisfactory would still be judged such at the present time.

Doppler Shift and A- and B-mode Ultrasound

The Doppler shift principle can be used to diagnose pregnancy using either a transducer mounted in a slim probe inserted into the rectum (Deas, 1977) where it can detect the flow of blood in the uterine artery, or a transducer applied externally to the abdomen to detect foetal heartbeats (Fukui, Kimura and Ono, 1984). In theory the detection of foetal heartbeat using the Doppler principle should provide a means of determining foetal numbers, but in practice the results of numerous trials have been disappointing. The use of intrarectal Doppler probes also gives cause for concern, as Tyrrell and Plant (1979) reported that about 50% of ewes so examined suffered some form of rectal damage.

The difficulty with this approach is that the heartbeats or pulses in the major blood vessels of one foetus can be picked up from different positions and angles of the transducer on the abdomen. To be certain that signals obtained from two different positions are coming from different foetuses, and not from two parts of the one foetus, it is necessary to measure the rate of the pulses and to determine that they are different from each other and from any maternal pulses which may also be detected. Such instruments have no range or depth discrimination, and consequently there is no way of knowing that an area of interest has not been fully examined as a result of gas or bone shadows or simply due to lack of sensitivity.

With a single-crystal transducer echoes coming from the interfaces or boundaries of the tissues through which the pulses of ultrasound are passing can be displayed on a screen in such a way that the depth of the interfaces beneath the skin surface are represented as distances in one dimension on the screen. This principle is used in A-mode scanning in which the transducer is generally placed on the bare area of the abdomen lateral to the udder and a single-point reflection of ultrasound from a particular depth, usually 9 cm or greater, is regarded as a positive diagnosis of pregnancy. The information obtained from such single-crystal transducers is obviously very limited, and many of the short-comings of Doppler shift instruments apply equally to A-mode scanners. Results from the use of this approach have been variable and generally leave much to be desired (Madel, 1983; Trapp and Slyter, 1983; Langford et al, 1984).

An extension of this approach is the B-mode scanner in which a single-crystal transducer is driven on a fixed rail and a series of one-dimensional signals from different positions are recorded photographically to give a two-dimensional representation of the tissues scanned. This type of instrument is capable of diagnosing pregnancy with a high degree of accuracy but results on the determination of foetal numbers have been disappointing (Lindahl, 1976).

Real-time Ultrasonic Scanning

Two types of real-time ultrasonic scanners can be used to diagnose pregnancy, estimate foetal age and determine foetal numbers in sheep, goats and cattle. The 'linear array' instrument has an array or row of crystals, usually 60-80 in number, arranged in a line some 10 cm long. The elements or groups of elements are fired sequentially and the signals or echoes are displayed as a series of close parallel lines on the screen. The 'sector' scanner has a small number of crystals which rotate or oscillate beneath a protective shield. The area of skin contact is small and the crystals are energized over a wide arc (up to 170°). The echoes are displayed as a series of diverging lines on the screen. The effect of either principle of use of ultrasound is to give a two-dimensional 'picture' or image of the underlying tissues from which it is possible to determine, for example, the shapes of various organs or parts such as the liver, heart or limbs. As the interfaces between tissues through which the ultrasound is travelling move, so movement is seen on the screen and this 'real-time' or live attribute of the displayed image is an important aid to image interpretation. Both types of instrument are used externally in sheep and goats and per rectum in cattle.

The above description is very simplistic and the instruments now available are in fact very sophisticated, incorporating microcomputers and other electronic devices to give images in varying shades of grey as well as black and white, and to improve generally the quality of the image displayed on the screen.

Acoustic Coupling

Ultrasound will not pass through air, and to ensure good transmission into the tissues, some form of coupling agent must be used between the transducer and the skin surface or rectal wall. Many different substances can be used, but proprietary gels are the most convenient and the most widely employed. With linear array instruments the gel is generally applied manually immediately before scanning, while with the only sector instrument commonly used for farm livestock, the gel is delivered from a pressurized container to the probe at the beginning of the scanning operation.

Linear Array Scanning

The use of linear array instruments to diagnose pregnancy and determine foetal numbers in sheep has been reported by White, Russel and Fowler (1984) and to diagnose pregnancy and estimate gestational age in cattle by White, Russel, Wright and Whyte (1985). The use of the technique in goats is essentially the same as in sheep.

Linear array instruments have a 'field of view' of approximately 8-10 cm wide by 20 cm deep, i.e. about 160-200 cm^2. In scanning sheep it is necessary to turn the ewe on its back, or at least into a reclined sitting attitude, to ensure that the whole of the area of potential interest is examined thoroughly by moving the transducer over the full width of the abdomen for a distance of some 20 cm in front of the udder. To achieve this the wool should first be clipped from that area. Some scanning operators now use this form of scanning without removing the wool, scanning only on the naturally bare areas of skin in that region. This practice may, however, lead to failure to detect some foetuses, particularly if multiple-bearing ewes are scanned at a more advanced stage of pregnancy.

Whether or not wool is clipped to extend the potential area of examination, it is essential that the transducer, normally held at right angles to the longitudinal axis of the ewe, is moved in a systematic pattern of searching from one side of the abdomen to the other, while maintaining good contact with the skin surface throughout. During the course of movement across the abdomen the angle at which the transducer is held against the skin is altered, making the ultrasound beam sweep or 'scan' areas of particular interest. In this manner the operator builds up a mental three-dimensional concept from the series of two-dimensional images viewed on the screen.

In scanning cows the transducer should ideally be placed in a polythene sleeve liberally covered with gel both inside and outside to ensure good acoustic coupling through the layer of polythene and to facilitate passage into the rectum. The transducer must be specially designed for rectal use and have the cable entering at the end and not on top. The transducer is introduced into the rectum in the operator's hand. The bladder is generally imaged as soon as the transducer has been passed over the pelvis, and the uterus can be viewed immediately anterior to the bladder. In the early stages of pregnancy the gravid uterine horn will lie to one side of the mid-line and can be imaged by a small change in the angle at which the transducer is held.

Sector Scanning

Although in theory any sector scanner can be used on farm livestock, there is only one instrument (the Oviscan 3, or its variant the Vetscan 2, BCF Technology Ltd, Livingston, Scotland) known to the author which has been specifically designed for this purpose and which is sufficiently robust to withstand the rigours of repeated use under the conditions usually encountered in practice. This instrument can be operated on an 85° or 170° sector and has a choice of depths of field ranging from 5 to 25 cm. Thus the area visualized at any one time can be up to 927 cm^2, in other words, approximately five times that of most linear array instruments. This means that in most cases, the whole of the area of potential interest can be viewed from a single point of contact. With sheep this allows the ewe to be scanned in a standing position and thus affords a considerable advantage in terms of labour and handling equipment over the normal linear array systems. Where pregnancy is more advanced, it may be necessary to move the transducer to the opposite side of the ewe, but in all cases it should be possible to image the entire uterus and its contents from contact with the skin on the naturally bare areas in front of the udder, thus obviating the need to clip wool from the abdomen.

The technique in cattle is essentially the same as for linear array instruments except that, because the sector probe's field of view is forwards as well as downwards, it is not necessary to introduce the probe as far into the rectum.

As in linear array scanning, a mental three-dimensional picture is built up by altering the angle at which the probe is held against the skin or rectal wall, thus sweeping or scanning across the entire width of the uterus.

Handling Systems

In scanning sheep with linear array instruments the ewe is either held in a sitting position or placed in a cradle like a low deckchair, as shown in Figure 1, so that the operator can move the transducer freely over the shorn area immediately in front of the udder and over the uterus.

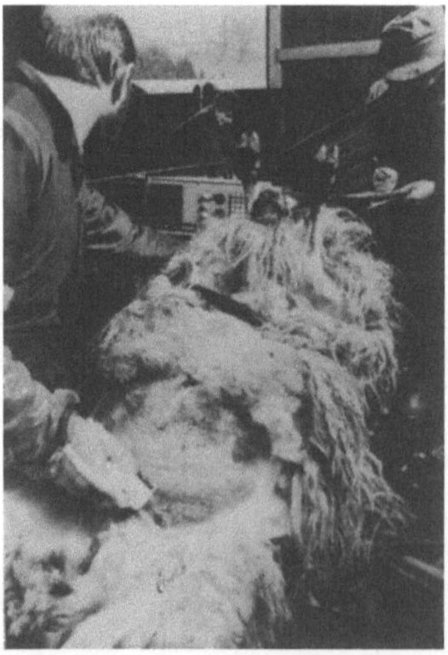

Fig. 1. Scanning a pregnant ewe restrained on a rotating handling system, using a linear-array instrument.

Catching the ewes, turning them up, shearing wool from the abdomen and restraining them for scanning is hard physical work, particularly with large ewes, and a team of three to four people in addition to the scanning operator, is required to ensure an efficient operation.

Mobile handling systems have been developed in which ewes in an elevated race are turned onto a cradle which rotates horizontally, first to a position for shearing and secondly to a further position for scanning. The ewes are presented for shearing and scanning at waist height which reduces fatigue and makes the operation more efficient and somewhat less labour demanding.

In sector scanning the ewes are generally restrained in a crate as illustrated in Figure 2. This has the considerable advantage that ewes enter and leave the crate themselves without having to be physically handled, thereby reducing the labour requirement and increasing the efficiency of the operation.

Goats tend to object violently and vocally to being turned on their backs, but can generally be scanned with a linear array instrument while restrained on their side on a flat surface such as a table or straw bale. Goats are also reluctant to enter the type of crate used for scanning ewes with the sector instrument, but can usually be restrained without difficulty by being held in a standing position as illustrated in Figure 3.

Fig. 2. Scanning a pregnant ewe restrained in a mobile crate, using a sector instrument

Fig. 3. Scanning a pregnant goat with a sector instrument

Cattle should be restrained for scanning in a crush or crate in which they are held by the neck. This prevents any sideways movement which could lead to damage to the instrument. It is important that there should be no gate or bar behind the cow as such a device could cause injury to the operator if the cow lies down during the scanning operation.

Stage of Pregnancy

The size of the sheep foetus at different stages of gestation can be judged from the illustration shown in Figure 4 in which the diameter of the coin is 2 cm. Goat foetuses are of a similar size at the stages of gestation shown. The skilled operator will be able to diagnose pregnancy by about 30 days post-conception from the identification of the fluid-filled uterus. At this stage of pregnancy, however, individual foetuses are too small to allow them to be identified with confidence. Cotyledons can be distinguished from about 40 days as white circular structures with hollow centres which appear and disappear as the angle of the transducer is altered. Individual foetuses can be identified by 45-50 days as white structures, frequently moving independently of the surrounding tissues. At that stage the head is about half the size of the body, the limbs are present as small buds, and the foetal length is about 4 cm.

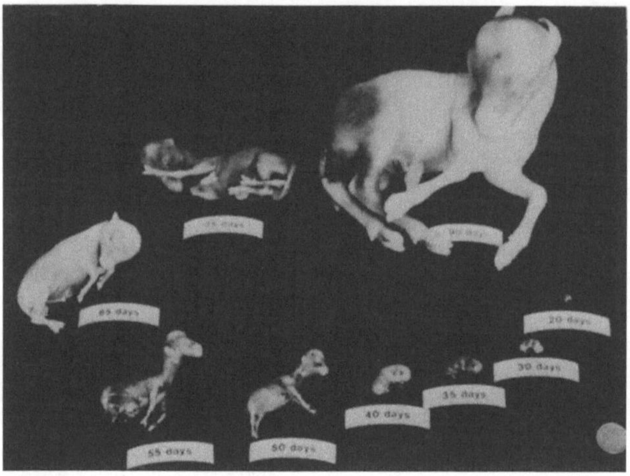

Fig. 4. Sheep feotuses at 20 to 90 days of gestational age.
The diameter of the coin is 2 cm

Foetal bones appear as bright white images from about 75 days. Ultrasound cannot pass through the calcified skeleton and thus black shadows appear underneath well-formed bony structures such as the skull and scapulae. Sections through the rib-cage show characteristic striated shadowing with black lines below each rib interspersed with white lines where ultrasound passes through the intercostal spaces. Photographs of images from pregnant ewes scanned with linear array and sector instruments are presented in Figs 5 and 6 respectively.

By 100 days images of individual foetuses more than fill the entire screen and only parts of each foetus can be viewed from any one position of the transducer. Shadows from the relatively large and dense bones make it difficult and often impossible to 'see through' one foetus to others which may be lying beyond. Foetuses can thus be counted accurately between 50 and 100 days of gestation. At later than 100 days the size of individual foetuses, the position of the uterus within the abdomen, and the shadows cast by the calcified foetal skulls and scapulae, combine to make the accurate determination of numbers difficult. With experience the recommended range of 50-100 days of gestation may be extended by 5 days at either end without affecting accuracy adversely.

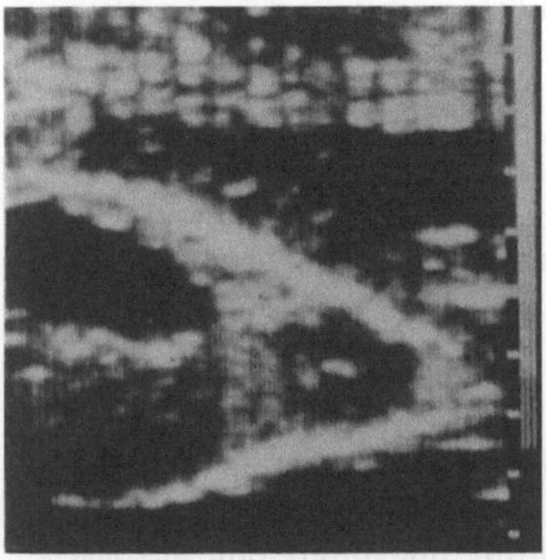

Fig. 5. Image from linear-array instrument showing thorax of foetus at about 100 days

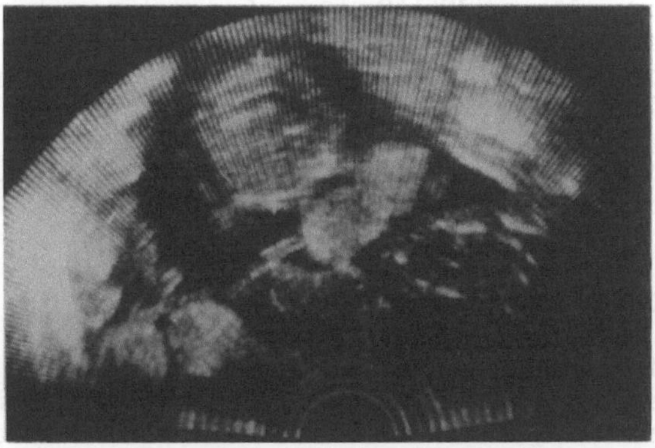

Fig. 6. Image from sector instrument of twin foetuses at 85 days of gestation.

Except where oestrus synchronization is practised, there will be a range in stage of gestation of ewes or goats in any flock or herd. The majority of animals will, however, be mated within two oestrus cycles, that is, within about 35-42 days. Thus scanning should be carried out after the latest mated animal is 45 days pregnant and before the first mated has reached 105 days. This gives an interval of between 3 and 4 weeks during which any one flock or herd can be scanned, i.e. from 80 to 105 days after the beginning of mating.

In cows pregnancy can be diagnosed with a high degree of accuracy from 30 days post-conception by the imaging of the fluid-filled uterus. At that stage the cotyledons are not developed and it can, on occasion, be difficult to locate the embryo. By 45-50 days the foetus has a distinct head and trunk, the latter with tail and limb buds, and by 70-80 days the cotyledons have developed their characteristic shape appearances. Photographs of images from pregnant cows scanned with linear array and sector instruments are presented in Figs 7 and 8 respectively.

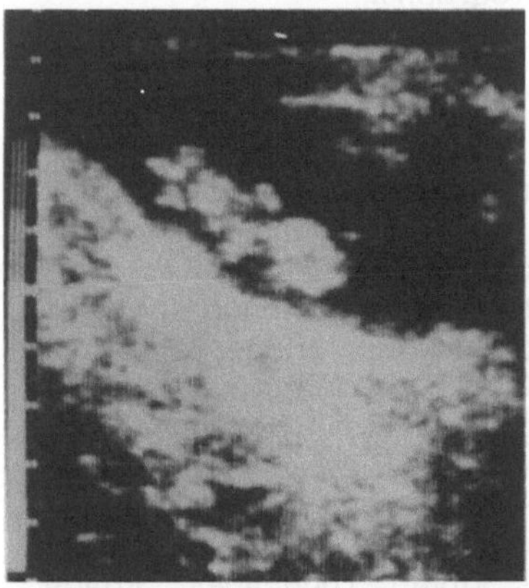

Fig. 7. Image from linear-array instrument of cattle foetus at 51 days of gestation

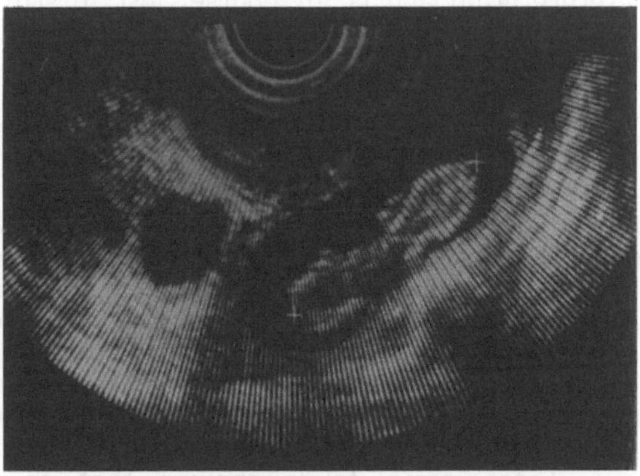

Fig. 8. Image from sector instrument of cattle foetus at 59 days of gestation

After 130-140 days of gestation, when the uterus descends into the abdomen, the foetus may lie beyond the 20-25 cm depth of the ultrasound beam, but pregnancy can still be accurately and confidently diagnosed from the fluid-filled part of the uterus and the cotyledons which are still visible. At this stage and later it is also possible to diagnose pregnancy by external examination, placing the transducer towards the lower surface of the abdomen immediately in front of the udder and to the right side of the mid-line.

Image Interpretation

The image viewed on the scanner screen is that of a section of the animal lying immediately beyond the transducer. As an aid to interpretation it can be useful to imagine that the animal has been cut across in the plane of the ultrasound beam and that the operator is looking into the cut surface. The image on the screen can be visualized as part of this cut surface, either rectangular or semicircular in shape, depending on the type of instrument used.

The areas of black, white and grey on the screen represent different tissues and their interfaces. Fluid-filled structures appear on the screen as black areas. Occasionally the bladder is seen as a black structure with well-defined smooth margins and lying in the midline position. The fluid-filled uterus of the pregnant animal also appears as a black area but, being convoluted in the early stages, may have an irregular outline and appear as several apparently discrete sacs, which, by moving the transducer, can be shown to be parts of one structure.

False or phantom pregnancies occur infrequently in goats and are readily identified by scanning, although difficult to diagnose by other means. In this condition the uterus is seen as a large fluid-filled area devoid of any structures such as cotyledons or membranes, and, of course, no foetus is present.

The movement of foetal limbs and bodies and the beating of foetal hearts seen in 'live' real-time images is an invaluable aid to interpretation and is, of course, lost in still photographs.

Image Quality

In the same way that air trapped beneath the skin surface and the transducer will prevent the transmission of ultrasound into tissue, so gas in the rumen and intestines in the path of the ultrasound beam will hinder the passage of ultrasound and cause poor image quality. Such conditions rarely occur in cattle and are most commonly found in ewes which have had recent access to roughages such as hay, silage or straw. Recent trials have shown that the adverse effects of intestinal gas on image quality can be minimized by withholding roughages from ewes for 8-10h before scanning. Starvation for more than 24h is not recommended as, in addition to possible harmful effects to the ewe, image quality appears to deteriorate again if food is withheld for longer periods.

These same trials showed that in ewes given roughages, image quality was markedly better in ewes scanned in the standing position with the sector instrument than in ewes scanned lying on their backs with a linear array instrument. This is considered to be more an effect of the disposition of the uterus in relation to the intestines than to the type of instrument per se, but is still a very considerable advantage in favour of the sector scanner.

Accuracy and Throughput

Accuracy is clearly a function of the scanning operator rather than the instrument. Given proper instruction most operators are capable of achieving very high levels of accuracy, particularly once they have gained some experience and, in certain cases, confidence. Most operators at the end of a 2 day course can scan sheep accurately at a rate of 25-30 ewes per hour. With experience the rate of scanning generally increases to perhaps 80-100 crossbred ewes per hour where all foetuses are to be counted. Where the flockmaster merely wishes to know whether ewes are carrying zero, one or more than one lamb, as is often the case with hill flocks in the UK, rates of 200 ewes or more per hour are commonly achieved.

The accuracy of diagnosis of pregnancy, if it can be guaranteed that no ram has been running with the ewes for the previous 45-50 days, should be virtually 100%. Certainly no ewe should be wrongly diagnosed as non-pregnant in these circumstances. Inevitably there will be instances of foetal death and resorption after scanning, and thus there will be a few cases where ewes correctly diagnosed as pregnant may fail to lamb. Where ewes are to be classified as carrying one or more than one foetus, a level of accuracy of 98% would generally be regarded as acceptable, although a number of operators can achieve consistently better results. Where the requirement is to count the exact number of foetuses in all ewes, the level of accuracy may fall to 96-97% where most ewes have twins or triplets and to lower levels where there is a significant frequency of even greater foetal numbers.

High levels of accuracy of pregnancy diagnosis can also be achieved in cows. In one trial carried out in The Macaulay Land Use Research Institute (Russel and Wright, 1986) 179 beef cows were scanned and 176 of the diagnoses agreed with the calving results. One cow which subsequently calved was diagnosed as non-pregnant and this was clearly an error on the part of the operator. Two cows which were diagnosed as pregnant failed to calve and since predictions of stage of gestation were also made in these cases it is possible that their diagnosis were correct at the time and that these foetuses subsequently died and were reabsorbed. The overall level of accuracy on the basis of calving results was therefore 98.3%. In another trial with 98 cows the accuracy was 99%.

The rate of scanning of cows is largely dependent on the standard of the handling facilities, but a throughput of 60 cows or more per hour can be readily achieved in practice.

Estimation of Foetal Age

Foetal age, which can be used to predict lambing, kidding or calving dates, can be estimated with a reasonable degree of precision from measurements made by scanning. Some instruments, such as the Vetscan 2, have the facility to 'freeze' the screen image, measure the size of foetal parts using electronic 'calipers', and predict from this measurement the age of the foetus. On instruments without this facility it is possible to construct a scale which can be used to assess foetal size on the screen image, or alternatively to make measurements on photographs of the screen image. Estimates of foetal age can then be read off graphs such as those shown in Figure 9 or calculated from specifically developed equations, such as presented for cattle in Table 1.

TABLE 1. Equations for predicting age (in days) of bovine foetuses from linear measurements made on scanning images (from White, Russel, Wright and Whyte, 1985)

Linear measurements (x) (cm)	Equation	RSD (days)
Head diameter	$37.7 + 45.23 \log_n x$	6.9
Head length	$25.7 + 40.38 \log_n x$	7.4
Trunk diameter	$39.7 + 37.21 \log_n x$	4.5
Cranium-rump length	$37.5 + 16.73 \log_n x$	4.5

No age-size relationships have been developed specifically for goats, but use of the information presented for sheep in Figure 9 is unlikely to lead to any serious errors in estimation of gestational age, and hence of kidding date. This would certainly be an adequate basis on which to

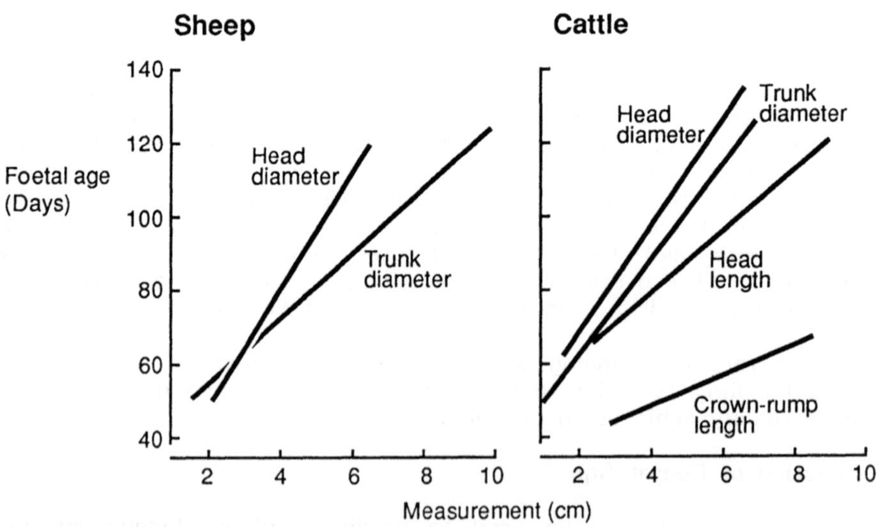

Fig. 9. Relationship between foetal age and size in sheep and cattle

divide a herd of pregnant goats into separate feeding groups for management in late pregnancy.

There is more information available on foetal age-size relationships and prediction of date of parturition in cattle than sheep. Only two measurements (head diameter and trunk diameter) are presented in Figure 9 for sheep, as it is relatively simple, by rotating the transducer to transform a longitudinal section into a transverse section. In rectal examinations of cattle, however, the transducer cannot be rotated through 90° and thus the option to choose any one of four measurement is given (Figure 9).

In trials with pregnant cows it was found that trunk diameter could be estimated more often than any of the other measurements. Crown-rump length, however, provided the most accurate estimate of age. In one study with 96 cows the difference between mean predicted and actual calving dates was 0.3 day; 58% of cows calved within 3 days of the predicted date and 90% within 10 days (Russel and Wright, 1986).

Uptake of Scanning in the United Kingdom

The first trials on the use of scanning in sheep in the UK were conducted in early 1983 (White, Russel and Fowler, 1984). Because it believed that the technique had something useful to offer to the sheep industry and that it was important that those operators pioneering the use of the technique should be as well trained as possible, the Hill Farming Research Organisation (now The Macaulay Land Use Research Institute) offered a series of courses of instruction on the use of real-time ultrasonic scanning to determine foetal numbers. In the first 'scanning season' in the 1983-84 winter, six commercial operators scanned a total of about 100,000 ewes (Russel, 1984). In the following 1984-85 winter some 60 commercial operators scanned about one million ewes (Russel, 1985), and it is estimated that in the recent 1987-88 winter some four million ewes, representing about 25% of the national flock, were scanned.

Most, but not all operators, have achieved a high degree of proficiency and many are now scanning between 30,000 and 40,000 ewes in a 10-week season, achieving, where flock sizes and physical conditions permit, throughputs of more than 1,000 ewes per day on occasion. The proficient operators now have as much work as they can deal with in the course of the scanning season, and those few who, for various reasons, have not achieved consistently good results, find it difficult to attract business. The signs are that scanning is now accepted as a significant aid to sheep management and production in the UK, and that the number of ewes scanned will continue to increase.

The major part of the demand for goat scanning stems from the large scale embryo transfer programmes carried out with cashmere and Angora goats, although dairy goat farmers and hobbyists are also seeking the services of skilled scanning operators. Currently some several thousand goats are scanned annually and the number is likely to increase substantially.

Although the scanning of cows was developed in the United Kingdom principally as an aid to the management of beef herds, an increasing number of dairy farmers are also making use of the technique for pregnancy diagnosis. As with goats, scanning is also regularly used to diagnose pregnancy and determine foetal numbers in embryo transfer programmes. At present the number of cows scanned annually is probably only about 15,000, but that figure is expected to increase rapidly over the next few years.

Acknowledgements

Grateful acknowledgement is made of the use of material from Pregnancy diagnosis and foetal number determination by I R White and A J F Russel in New Techniques in sheep production edited by Marai & Owen and published by Butterworths, London, 1987.

References

BEACH, A.D. (1982) 'Multiple pregnancy diagnosis in sheep using pulsed video-fluoroscopy and a semiconductor video store', New Zealand Journal of Experimental Agriculture, 10, 291-295.

DEAS, D.W. (1977) 'Pregnancy diagnosis in the ewe by an ultrasonic rectal probe', Veterinary Record, 101, 113-115.

FUKUI, Y., KIMURA, T. and ONO, H. (1984) 'Multiple pregnancy diagnosis in sheep using an ultrasonic method', Veterinary Record, 14, 145.

HULET, C.V. (1973) 'Determining foetal numbers in pregnant ewes', Journal of Animal Science, 36, 325-330.

LANGFORD, G.A., SHRESTHA, J.N.B., FISER, P.S., AINSWORTH, L., HEANEY, D.P. and MARCUS, G.J. (1984) 'Improved diagnostic accuracy by repetitive ultrasonic pregnancy testing in sheep', Theriogenology, 21, 691-698.

LINDAHL, I.L. (1976) 'Pregnancy diagnosis in ewes by ultrasonic scanning', Journal of Animal Science, 43, 1135.

MADEL, A.J. (1983) 'Detection of pregnancy in ewe lambs by A-mode ultrasound', Veterinary Record, 112, 11-12.

RICHARDSON, C. (1972) 'Pregnancy diagnosis in the ewe: a review' Veterinary Record, 90, 164-275.

ROBERTS, S.J. (1971) ' Veterinary Obstetrics and Genital Diseases. Ithaca, New York.

RUSSEL, A.J.F. (1984) 'Ultrasonic scanning', The Sheep Farmer, 4(1), 30-31.

RUSSEL, A.J.F. (1985) 'Sheep scanning', The Sheep Farmer, 5(2), 31-32.

RUSSEL, A.J.F., DONEY, J.M. and REID, J.L. (1967) 'Energy requirements of the pregnant ewe', Journal of Agricultural Science, Cambridge, 68, 359-363.

RUSSEL, A.J.F. and WRIGHT, I.A. (1986) 'Scanning: an aid to management in suckler herds', Science & Quality Beef Production, Agricultural and Food Research Council, London.

THOMSON, W.A. and AITKEN, F.C. (1959) 'Diet in relation to reproduction and viability of the young', Part II: Sheep. Commonwealth Agricultural Bureaux, England.

TRAPP, M.J. and SLYTER, A.L. (1983) 'Pregnancy diagnosis in the ewe', Journal of Animal Science, 57, 1-5.

TURNER, M.J. and HINDSON, J.C. (1975) 'An assessment of a method of manual pregnancy diagnosis in the ewe', Veterinary Record, 96, 56-58.

TYRRELL, R.N. and PLANT, J.W. (1979) 'Rectal damage in ewes following pregnancy diagnosis by rectal-abdominal palpation', Journal of Animal Science, 48, 348-350.

WENHAM, G. and ROBINSON, J.J. (1972) 'Radiographic pregnancy diagnosis in sheep', Journal of Agricultural Science, Cambridge, 78, 233-238.

WHITE, I.R., RUSSEL, A.J.F. and FOWLER, D.G. (1984) 'Real-time scanning in the diagnosis of pregnancy and the determination of foetal numbers in sheep', Veterinary Record, 115, 140-143.

WHITE, I.R., RUSSEL, A.J.F., WRIGHT, I.A. and WHYTE, T.K. (1985) 'Real-time ultrasonic scanning in the diagnosis of pregnancy and the estimation of gestational age in cattle', Veterinary Record, 117, 5-8.

WILSON, I.A.N. (1981) 'Ovine pregnancy diagnosis', Grassland Research Institute Technical Report No. 28.

THE DIAGNOSIS OF PREGNANCY AND PSEUDOPREGNANCY, AND THE DETERMINATION OF FOETAL NUMBERS OF GOATS, BY MEANS OF REAL-TIME ULTRASOUND SCANNING.

M.C. Lavoir and M.A.M. Taverne
Department of Herd Health and Reproduction
State University of Utrecht
Yalelaan 7
3584 CL Utrecht, The Netherlands

INTRODUCTION

The goat is a seasonal breeder and usually gives birth once a year. The earliest sign of pregnancy after mating or artificial insemination (AI) is the absence of oestrus. This can only be detected during the oestrous season, but even then not all non-pregnant goats will return to oestrus. Another complicating factor is that some pregnant goats may show signs of false heat. For a more evenly distributed milk production through the year, farmers sometimes want their does to give birth in the autumn. This means that oestrus and ovulation have to be induced by hormonal treatment during the anoestrous season. Unfortunately, the pregnancy rate after mating during the anoestrous season is always lower than that during the oestrous season (Corteel et al. 1982). It is assumed that mating after hormonal treatment will result in more pseudopregnancies (Mizinga et al. 1984).

An early test for pregnancy is essential should a farmer want his goats to give birth more than once a year. Testing would be more helpful if it could also be used to predict the expected number of kids. Subsequent adjustment of food allotments during late pregnancy should then reduce the number of overweight single kids, the number of small sized multiples and the incidence of pregnancy toxemia.

Since we began using real-time ultrasound scanning for pregnancy diagnosis and foetal counting, several pseudopregnancies have been diagnosed (Pieterse et al. 1986). In pseudopregnant goats a so-called hydrometra exists; a pathological condition of the uterus in which aseptic fluid accumulates in the presence of a persistant corpus luteum. Hydrometra was detected in goats which had been presented for pregnancy diagnosis, or for various disorders such as permanent anoestrus and failure to return to oestrus following sterile mating. Treatment of a hydrometra consists of a single injection of prostaglandin F2 . Within two days after treatment the aseptic fluid will be discharged - this is also called a cloudburst - and depending on the time of the year, breeding may be resumed.

Until recently there was no practical method for the diagnosis of hydrometra in the goat. While oestrone sulphate levels in milk or plasma may be used to diagnose pregnancy, they are probably not useful in differentiating between a hydrometra and a non-pregnant animal. The doppler test and radiography are two techniques which may also be used to diagnose pregnancy. However, the accuracy of the doppler technique varies considerably (Berkley 1982; Wani 1981; Ott et al. 1981), while

M. M. Taverne and A. H. Willemse (eds.), Diagnostic Ultrasound and Animal Reproduction, 89–96.

radiography can only be used in well equipped centres after day 70 of gestation (Barker et al. 1967). Blood progesteron measurements, A-mode ultrasound, and rectal and abdominal palpation cannot be used to distinguish pregnancy from pseudopregnancy (Holdsworth et al. 1979; Ott et al. 1981; Berkley 1982).

This paper describes experiments aimed at studying the accuracy of pregnancy diagnosis, the accuracy of determining foetal numbers, and the incidence of hydrometra in goats by means of real-time ultrasonography.

MATERIALS AND METHODS

Animals

In 1985 we began to predict foetal numbers in naturally mated goats at an experimental farm in the Netherlands (IVO, Zeist). The animals of the same herd were also used in 1986 and 1987. The investigators had experience in the prediction of foetal numbers in sheep.

During a field trial in 1986 goats were tested for pregnancy with ultrasonography by an inexperienced operator. The herds used in this study were involved in an AI experiment. We participated in this experiment because it was expected that the pregnancy rate would be lower than in a herd served by a buck. This was an excellent opportunity to determine the specificity of the test on a larger group of goats.

The stage of pregnancy on the day of scanning was calculated on the basis of the recorded day of insemination/mating. In both studies nulliparous as well as pluriparous does of the Saanen breed were used.

Equipment

A portable scanner type 400 (Pie Medical, Maastricht, the Netherlands) was used to predict foetal numbers. The scanner was provided with a 3.0 and a 5.0 MHz linear-array transducer.

During the field trial the same scanner was used as well as an ADR 4000SL scanner provided with a 5.5 MHz sector transducer (Scientific Medical Systems, Dordrecht).

Scanning procedure

All animals were scanned in a standing position. A section of the right ventral abdominal wall was shaved. The animal was restrained by an assistent during scanning.

Positive pregnancy diagnosis was based, depending on the stage of gestation, on the recognition of a fluid filled uterus, placentomes, foetal structures such as head, thorax and limbs, or foetal body movements. The confirmation of pregnancy was based upon kidding results 145-155 days after insemination.

To determine the foetal number, a scanning period of 2-3 minutes was usually necessary. Screening of the entire uterus was attempted by slowly turning the transducer along its longitudinal axis. Independent movement is an important criterion for distinguishing the individual foetus. Shaking the abdominal wall during the scanning procedure sometimes induced foetal movements which facilitated the count.

Visualization of a foetus or of foetal membranes is not always possible before day 35. The ultrasonic appearance of a hydrometra and of an early pregnant uterus may therefore be very similar. To differentiate pregnancy from hydrometra, scanning should be performed at least 40 days after mating. At this stage in the pseudopregnant animals no foetal parts and placentomes are present and only fluid-filled compartments, seperated by thin tissue walls, are visible. These thin tissue walls are formed by the fluid-filled uterine horns curving under the corpus uteri Figure 1. The ultrasound picture shows various sections through these curved uterine horns. The tissue walls may be identified as two adjacent parts of the uterine wall (Pieterse et al. 1986).

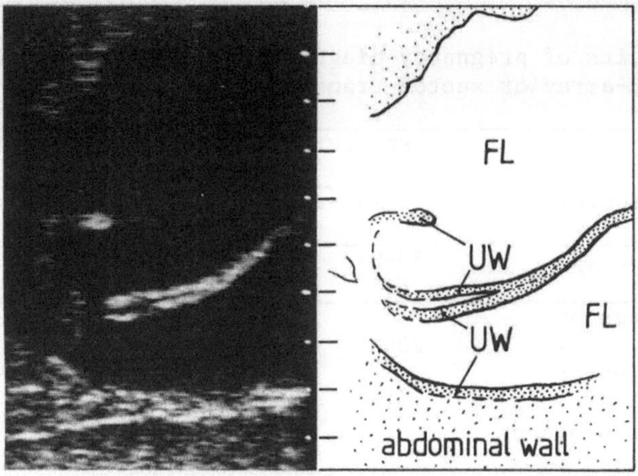

Figure 1. Ultrasonographic picture of hydrometra in the goat. Fluid-filled (FL) non-echogenic compartments are seperated by tissue walls formed by two adjacent uterine walls (UW).

Processing of the data

Sensitivity, specificity, as well as positive and negative predictive values of the pregnancy diagnoses were calculated. The sensitivity is the ability of the test to identify the pregnant animals, while the specificity is the ability of the test to identify the non-pregnant animals. The predictive value of the test gives an indication of the certainty with which statements can be made about the presence (positive predictive value) or absence (negative predictive value) of pregnancy.

The sensitivity of the foetal count is the ability of the test to identify 1, 2 or 3 and more kids. The predictive value indicates the certainty with which statements can be made about the presence of 1, 2, or 3 or more kids.

RESULTS

Pregnancy examnination was performed in 632 animals of which 254 appeared to be open at the expected time of kidding. In Table 1 the results of pregnancy testing are arranged in three groups according to the stage of gestation during scanning and the type of scanner that was used. Specificity in all groups is more than 95%. The sensitivity is close to 100% while the positive and negative predictive values are above 95% in all groups. Performing the test between day 25 and 30 (7 animals; not included in table 1) resulted in a sensitivity of 50%, due to an equal number of 3 incorrect diagnoses "non-pregnant" and 3 correct diagnoses pregnant.

TABLE 1. Results of pregnancy diagnosis by means of ultrasonography with a linear-array or sector transducer

	Exp. farm	Field trial	
type of transducer	linear-array	linear-array	sector
days of pregnancy	30-91	50-70	40-50
diagnosis pregnant correct (a)	204	105	68
diagnosis pregnant incorrect (b)	1	3	3
diagnosis not pregnant correct (c)	34*	65	148
diagnosis not pregnant incorrect (d)	1**	0	0
sensitivity (a/a+d)	99.5%	100%	100%
specificity (c/c+b)	97.1%	95.6%	98%
positive predictive value (a/a+b)	99.5%	97.2%	95.8%
negative predictive value (c/c+d)	97.1%	100%	100%

* including 5 does with hydrometra

** scanning took place on day 32 of gestation

After day 30 hydrometra was diagnosed in 5 of the 240 goats at the experimental farm. These were subsequently confirmed by the discharge of only uterine fluids after treatment with prostaglandins. During the field trial the inexperienced operator did not diagnose animals with hydrometra.

Figure 1 and Table 2 summarize the results of the fetal counts. The results are divided into two groups. The first group represents the scans made between days 40 and 70, the second group represents those made between 71 and 91 days. Sensitivity and predictive values for the first group are 90.5% and 70.4% respectively for singles, 88.1% and 98.3% for twins and 91.4% and 88.9% for triplets or more. When only two categories are used (singles and twins or more) these figures are 92.2% and 97.9% for respectively twins or more. The results of the second, smaller group (day 71-91) clearly show that the prediction of foetal numbers at this stage has a low accuracy.

		Days 40-70 (n=140) Kids born		
		1	2	3
fetuses counted during scanning	1	19	5	3
	2	1	59	0
	3	1	3	32

		Days 71-91 (n=23) Kids born		
		1	2	3
fetuses counted during scanning	1	1	1	3
	2	0	6	11
	3	0	1	0

Figure 1. Comparison of the predicted number of fetuses with the number of kids born.

TABLE 2. Accuracy of counting fetal numbers in goats by means of transabdominal ultrasonography.

	scanning at days	
	40-70(n=140)	71-91(n=23)
sensitivity, 1 kid	90.5%	100.0%
predictive value, 1 kid	70.4%	20.0%
sensitivity, 2 kids	88.1%	75.0%
positive pred. value, 2 kids	98.3%	35.3%
sensitivity, $\geq$ 2 kids	92.2%	81.8%
predictive value, $\geq$ 2 kids	97.9%	100.0%
sensitivity, 3 kids	91.4%	0%
predictive value, 3 kids	88.9%	0%

DISCUSSION

Our data showed that pregnancy testing with transabdominal ultrasonography before day 30 results too many false negative diagnoses. It is our impression that adequate visualization by transabdominal scanning is hampered by an almost intrapelvic location of the uterus at this early stage of gestation.

In the field trial there were six false positives which we believe were not fully accounted for by fetal death or undetected abortions. We strongly suspect that at least some of the six false positives were in fact animals with hydrometra, because no distinction was made between pregnancy and pseudopregnancy by the inexperienced researcher. The one false positive diagnosis made at the experimental farm occured in a goat that did not kid; however this animal had a non-viable fetus during scanning. Due to these "failures" specificity did not reach 100%.

There appeared to be no difference between the results obtained by using a linear-array or a sector transducer. Contrary to expectation, the overall accuracy of intrarectal real-time echography performed between day 20 and 60 was lower (80-90%, Botero-Herrera et al. 1984) than the accuracy obtained in our study by transabdominal scanning (95-100%).

100% accuracy is claimed for the diagnosis of pregnancy using the oestron sulphate measurement in milk (Davies et al. 1983). This high accuracy level may be attributed to the fact that oestron sulphate is produced by the conceptus. Because this test differentiates only between pregnant and non-pregnant animals, the pseudo-pregnant goats with a hydrometra will probably be diagnosed simply as non-pregnant. In order to treat goats with a hydrometra, all non-pregnant goats will therefore have to be treated with prostaglandins.

The ultrasonic diagnosis of hydrometra is based upon the recognition of uterine fluid in the absence of foetuses and placentomes (Pieterse et al. 1986). Smith (1986) describes the fluid of a hydrometra as being clear to cloudy. We saw only hydrometra containing clear fluid as recognized by the appearance totally black shadows on the ultrasonic images. Cloudy fluid, characterized by a snowy appearance on the ultrasonic image, is associated with other pathological conditions such as fetal death during early pregnancy or pyometra post-partum (Taverne, unpublished data).

In the group of animals at the experimental farm scanned after day 30 the incidence of pseudopregnancy was 2.1%; 1 out of 65 goats in 1985, 3 out of 89 in 1986 and 1 out of 86 in 1987. Holdsworth (1979) found 4 animals with hydrometra in a group of 98 goats. Botero-Herrera (1984) did not report cases of hydrometra. The false positive diagnoses in the latter study were all attributed to embryonic death. In a recent field trial on 3 different farms in the Netherlands, 8 pseudopregnant animals were found in a total number of 262 goats (= 3.1%). The observation of Mizinga et al. (1984), that pseudopregnancy occur more often after induced ovulations, has recently been confirmed by the findings of Hesseling (personal communication 1989) who diagnosed 5 pseudopregnancies in a group of 10 goats which were mated after treatment with MAP-sponges and PMSG. This indicates that the anoestrus caused by pseudopregnancy is sometimes a serious problem, especially when superovulation has been induced.

As in the case of sheep (Taverne et al. 1985), the best period for counting foetal numbers in goats, using transabdominal scanning, appears to be between days 40 and 70. Sensitivity and predictive values calculated between days 71 and 91 are clearly lower, although only a small number of animals was examined at this stage. Sometimes underestimations are made, especially in does carrying triplets. The foetuses are concealed behind each other and the penetration depth of the 3.0 MHz transducer is too small to visualize the uterine contents at a distance greater than 20 cm from the abdominal surface. Because the uterus descends into the abdomen, the intestines also complicate the picture. Over estimations are rare. Based on our experience with foetal counts in sheep, we suspect that in these few cases foetal death took place at some stage after pregnancy testing. If a differentiation has to be made between singles, twins, and three or more, the sensitivity of our method is rather low. However, if the differentiation is only made between single and two or more kids, the sensitivity and predictive values of our tests are satisfactory. According to Botero-Herrera (1984), the determination of foetal numbers by transabdominal scanning is more accurate than by intra-rectal scanning.

We conclude that trans-abdominal real-time ultrasound scanning is an easily applied test with a high accuracy both for pregnancy diagnosis and for determining foetal numbers in goats. In addition, it enables early recognition of pseudopregnant animals with hydrometra.

REFERENCES

Barker, C.A.V. and Cawley, A.J. (1967) "Radiographic detection of fetal numbers in goats", Can. Vet. J. 8, 59-61.

Berkley, P.M. (1982) "Pregnancy testing of goats using ultarasonics", Goat Vet. Soc. J. 3, 13-14.

Botero-Herrera, O., Gonzâlez-Stagnaro, C., Poulin, N. and Cognie, Y. (1984) "Diagnostico precoz de la gestacion en las cabras y ovejas utilizando la echografia de ultrasonido por via rectal", Proceedings 10th Int. Congr. Anim. Reprod. and A.I., Urbana, Illinios, volume II, 79.

Corteel, J.M., Gonzalez, C. and Nunes, J.F. (1982) "Research and development in the control of reproduction", Proceedings 3rd Int. Congr. on Goat Production and Disease, Tucson, Arizona, pp. 584-591.

Davis, J. and Chaplin, V.M. (1983) "A caprine pregnancy test based upon measurement of oestrone sulphate in milk", Proceedings 3rd Int. Symp. World Assoc. Vet. Lab. Diagn., Ames, Iowa, 1, 119-125.

Holdsworth, R.J. and Davis, J. (1979) " Measurement of progesterone in goats milk: an early pregnancy test", Vet. Rec. 105, 535.

Mizinga, K.M. and Verma, O.P. (1984) "LHRH-induced ovulation and fertility of anoestrus goats", Theriogenology 21, 435-446.

Ott, R.S., Brown, W.F., Lock, T.F., Memon. M.A. and Stowater, J.L. (1981) "A comparison of intrarectal doppler and rectal abdominal palpation for pregnancy testing in goats", J. Am. Vet. Med. Assoc. 178, 730-731.

Pieterse, M.C. and Taverne, M.A.M. (1986) " Hydrometra in goats: Diagnosis with real-time ultrasound and treatment with prostaglandins or oxytocin", Theriogenology 26, 813-821.

Smith, M.C. (1986) "Anoestrus, Pseudopregnancy and cystic Follicles", in: D.A. Morrow (ed.), Current Therapy in Theriogenology 2, W.B Saunders, Philadelphia, pp. 585-586.

Taverne, M.A.M., Lavoir, M.C., van Oord, R. and van der Weyden, G.C. (1985) "Accuracy of pregnancy diagnosis and prediction of fetal numbers in sheep with linear-array real-time ultrasound scanning", Vet. Quaterly 7, 256-263.

Wani, G.M. (1981) "Ultrasonic pregnancy diagnosis in sheep and goats: A review ", World Review of Anim. Prod. 17, 43-48.

THE USE OF LINEAR ARRAY REAL-TIME ULTRASONOGRAPHY FOR PREGNANCY DIAGNOSIS IN PIGS.

M.A.M. Taverne, Department of Herd Health and Reproduction, State University of Utrecht, Yalelaan 7, 3584 CL Utrecht, The Netherlands

INTRODUCTION

Early detection of open females provides an economic basis for pregnancy tests on pig farms. Oestrus detection alone will not enable the identification of all nonpregnant females because nonpregnant pigs may fail to show subsequent oestrous behaviour due to prolonged cycle lengths, silent oestrus, anoestrus or cystic ovaries. On well managed pig farms with farrowing rates higher than 80%, it can be expected that 5-10% of the animals which do not return to oestrus at 3 weeks post mating or insemination will be nonpregnant. This means that the costs of any pregnancy "test" other than oestrus detection, rest upon a relatively small portion of all anoestrous sows that have to be tested. Therefore, a pregnancy test should be highly relaible in order to be profitable. It has also been (theoretically) calculated (van der Steen, 1985) that a negative pregnancy test arround day 35 of gestation which will not be followed by therapeutic measures such as induction of oestrus, is only of economic interest when farrowing rate are lower than 80%; it may, however, lead to a reduction in net return (returns minus costs, labour included) at higher farrowing rates. So both the accuracy of pregnancy diagnosis and decision making on the basis of this test are of outmost importance.

Since 1983 several reports have appeared in the literature describing the application of two-dimensional real-time ultrasonography for early pregnancy detection in the pig. These studies indicated that this technique is highly reliable and provides an instantaneous answer so that therapeutic measures or culling can take place immediately after testing.

This contribution will review the presently available information on this subject with special reference to the methodology and the accuracies that were obtained with tests on the farm.

METHOD OF INVESTIGATION

In all the studies published so far, linear-array transducers have been used. This is mainly for two reasons: 1) the pregnant uterus, even at an early stage of gestation, rests upon the ventral abdominal wall so that the structures are close to the transducer and visualization should be possible; 2) the greater contact surface of the linear-array transducer as apposed to that of the sector transducer, guaranties better visualization, especially when the animal makes body movements. Although the ultrasound frequency of the transducers that were used, ranged from 2 MHz to 9 MHz, most studies were performed with a 3.5 or 5 MHz frequency (Martinat-Botté et al, 1985, 1988; Inaba et al, 1983, Taverne et al,

M. M. Taverne and A. H. Willemse (eds.), Diagnostic Ultrasound and Animal Reproduction, 97–103.

1985). It is our experience that the resolution of the 3.5 MHz transducer is good enough for this diagnostic application and enables a penetration depth of 17-20 cm.

Very little attention has been paid by manufacturers to adapting the design of scanners for convenient use on farms. Although several scanners are supposed to be portable, they are still heavy and very difficult to handle for periods longer than 15 minutes while climbing or walking from one cage to another. Battery-operated scanners were not available in any of the published studies.

Pregnancy testing takes place while the animal is standing. In one study (Botero et al, 1986) a comparison was made between tests performed with animals which were either tethered, or in a retention box of a closed or semi-outdoor building. Higher accuracies were obtained in the sows tethered by neck or around the shoulder; scanning under these conditions are less hampered by movements of the sow. Placement of the transducer, provided with coupling gel, is either on the left or right paramedian ventral abdominal wall, just above the row of teats and immediately cranial to the hindleg. The long axis of the transducer remains parallel to that of the sow while it is slowly moved in the cranial and caudal directions (figure 1).

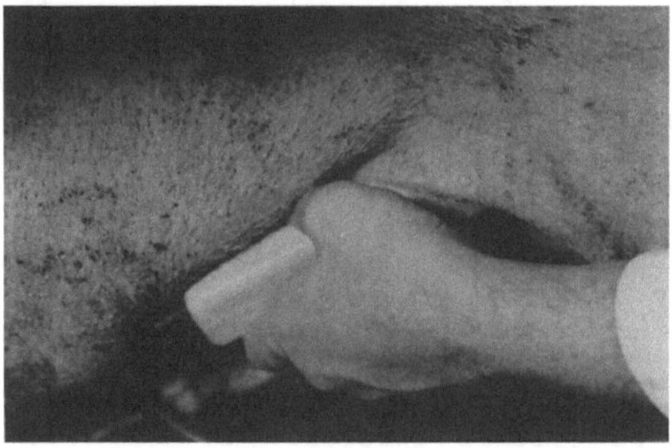

Figure 1. The transducer is placed on the ventro-lateral abdominal wall, just dorsal to the row of teats and cranial to the hindleg.

There appears to be no subtantial increase in accuracy when scanning takes place both on the left and the right side of the animal (Martinat-Botté et al, 1988). Experience of the operators with the technique of scanning and interpretation of the images clearly improved the results of testing (Botero et al, 1986) although it is suggested by one study (Martinat-Botté et al, 1988) that experience is already acquired when more than 200 investigations have been performed.

Only one study reports on the use of transrectal scanning (Cartee et al, 1985) but the authors concluded that transabdominal scans were much easier and much less traumatic to the sows.

Between days 20 and 30 of gestation there is a rapid increase in volume of allantoic fluid associated with each of the many conceptuses. At this stage the size of the embryo itself and the volume of the amniotic fluid are still relatively small (Knight et al, 1977).

Therefore scanning the sow at this early phase of gestation implicates that a positive diagnosis is based on the recognition of the presence of fluid in the uterus. Because the convoluted uterus will be sectioned at several places by the plane of the ultrasound beam, a pregnancy will be recognized by the presence of several non-echogenic (black) areas, at the cranioventral side of the bladder (figure 2).

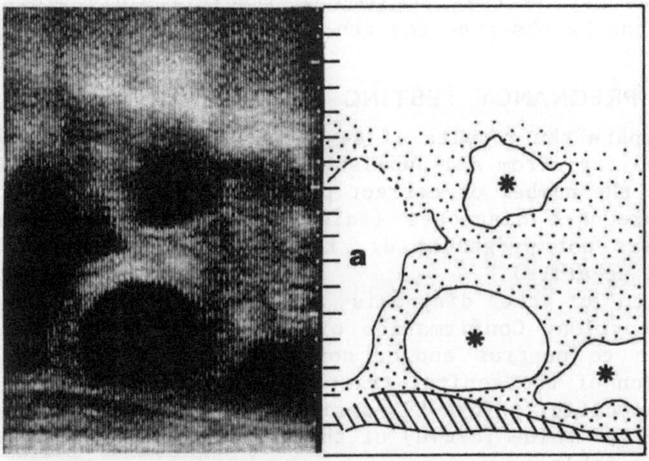

Figure 2. Scan of a sow at the 26th day after mating. Three sections through the allantoic-fluid filled uterus (*) can be seen cranial to the bladder(a). The abdominal wall is at the lower end of the picture (scale in cm).

These areas are seperated by more echogenic tissue consisting of two apposing uterine walls. By turning the transducer along its longitudial axis, the small embryo can sometimes be seen (figure 3).

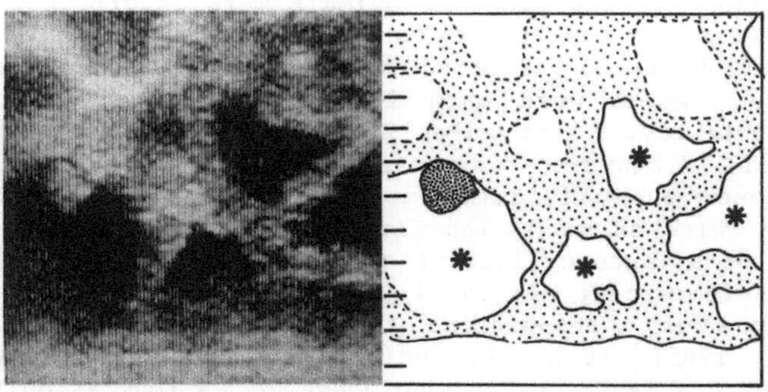

Figure 3. Scanning picture of a sow at day 30 of gestation. Within one of the sections through the allantoic-fluid filled uterus (*) an embryo is visible (stippled). The abdominal wall is at the lower end of the picture.

However under farm conditions a positive diagnosis is usually based only on the recognition of fluid in the uterus. The bladder will not interfere with the diagnosis because urine in the bladder will appear as a single, caudally situated non-echographic area on the monitor (figure 2). The increase in volume of allantoic fluid between days 20 and 30 of pregnancy is reflected by the increase in the mean diameter of the non-echographic areas seen during scanning (Martinat-Botté et al, 1988). It is only after day 29 that the embryo itself rapidly grows and that the heart beat can be observed for the first time by ultrasonography.

ACCURACY OF PREGNANCY TESTING

In order to compare the results of several field experiments which have been reported so far, from all published data the following numbers were calculated: a). the number of correct pregnant diagnoses; b). the number of incorrect pregnant diagnoses (false positive); c). the number of correct diagnoses not pregnant; d). the number of incorrect diagnoses pregnant (false negative).

Confirmation of the diagnosis pregnant was based on observed farrowing or abortion. Confirmation of the diagnosis not pregnant was based on return to oestrus and/or non-farrowing on the expected time and/or inspection of the genital tract of slaughtered animals. Sensitivity (a/a+d), specificity (c/c+b), positive predictive value (a/a+b) and negative predictive value (c/c+d) of the different tests were calculated and are given in table 1.

In all tests 9365 animals have been investigated; 8145 of these proved to be pregnant and 1220 were not, according to the above described criteria. However it seems appropriate to distinguish those studies in which tests were also performed on days 18, 19, 20 and 21 from the other ones because at this rather early stage, selection of nonpregnant animals by means of oestrus detection could not yet have taken place. Indeed Botero et al (1986) showed that more than 50% of the sows diagnosed not pregnant between days 18 and 25 after insemination returned to oestrus within one week after ultrasonography.

Evaluating these early tests (marked with * in table 1), it appears that 4.3% (59/1385) of the pregnant animals were incorrectly diagnosed and 41.3% (104/252) of the nonpregnant sows were incorrectly diagnosed as pregnant. For the tests performed at a later stage (marked with ** in table 1) these figures were much lower: 0.45% (31/6769) and 18.7% (181/968) respectively. So it makes no sense to apply ultrasonography before the first "return-to-oestrus-period" has passed because false negative diagnoses occur 10 times more frequently and false positive ones more than twice as much. The first are most likely caused by the smaller volume of allantoic fluid at this early stage especially in the smaller litters (Botero et al, 1986), the latter by embryonic mortality and the presence of intra-uterine fluid other than allantoic fluid (Taverne et al, 1985). It is very unlikely that specificity (c/b+c) of the test can reach a level much higher than 90% as long as only observed farrowing or abortion are used for confirmation of the diagnosis pregnant. In this respect it would be interesting to combine ultrasonography with determination of oestrone sulphate in either maternal plasma (Saba et al, 1981; Atkinson et al, 1986), urine, or faeces (Choi et al, 1987).

DATA FROM

	Inaba et al 1983		Martinat-Botté et al 1985		Taverne et al 1985	Martinat-Botté et al 1988		
	days p.c. /	p.i.	days p.c. /	p.i.	days p.c. / p.i.	days	p.c. / p.i.	
	19–22	23	18–23	24	24–32	18–21	22–30	31
	(n=42)	(n=80)	(n=1044)	(n=1213)	(n=881)	(n=551)	(n=2474)	(n=3080)
	*	**	*	**	**	*	**	**
DIAGN. + CORR.	24	64	873	1080	785	429	2120	2680
DIAGN. + INCORR.	0	0	71	57	9	33	66	49
DIAGN. - CORR.	7	16	78	73	87	63	277	334
DIAGN. - INCORR.	11	0	22	3	0	26	11	17
SENSITIVITY %	68.6	100.0	97.5	99.7	100.0	94.3	99.5	99.4
SPECIFICITY %	100.0	100.0	52.3	56.1	90.6	65.6	80.8	87.2
+PREDICTIVE VALUE %	100.0	100.0	92.4	94.9	98.9	92.8	97.0	98.2
-PREDICTIVE VALUE %	38.9	100.0	78.0	96.0	100.0	70.8	96.2	95.1

Table 1. A comparison between the accuracies of several pregnancy tests performed by means of two dimensional ultrasonography (asterisks are explained in the text).

Very little information is presently available on the influence of parity and breed of the animals on the accuracy of ultrasonographic pregnancy diagnosis. The significantly lower negative predictive value in pluriparous sows and in animals of the Large White breed as compared with that in gilts and in animals of crossbreeds (Martinat-Botté et al, 1988) awaits confirmation and explanation.

TECHNOLOGY ASSESSMENT

Like with the introduction of each new diagnostic technique, the last phase should consist of an evaluation of economic results on farms using the technique. In this respect it is amazing that hardly any published data is yet available on this topic for any of the methods presently used on pig farms. Specificity of routine A-mode testing between days 31 and 37 performed by farm employes in one comparative study (Taverne et al, 1985) was only 55.8% and one would expect a decrease of net return rather than an improvement of economic results compared with the use of oestrus detection alone (van der Steen, 1985).

In a french study (Matinat-Botté et al, 1988) two major changes have been noted on farms where a period of 12 months, during which only oestrus detection was used, was followed by a year in which oestrus detection and ultrasonography were applied: 1) there was an increase in the number of animals seen in oestrus between 38 and 45 days after mating/A.I. from 9.1% during the first to 18.3% during the second year; 2) of the culled animals the mean interval between mating and culling decreased from 96 to 46 days. Further studies on these topies will have to show wether the use of two-dimensional ultrasonography will be beneficial or not. In the meantime it should be noted that the same equipment can possibly be used for the diagnosis of reproductive disorders in sows (Botero et al, 1986; Madec et al, 1988) as well as for measurements of back-fat thickness.

REFERENCES

Atkinson, S., Buddle, J.R., Williamson, P., Hawkins, C.D. and Wilson, R.H. (1986) "A comparison between plasma oestrone sulphate concentration and Doppler ultrasound as methods for pregnancy diagnosis in sows", Theriogenology 26, 483-490.

Botero, O., Martinat-Botté, F. and Bariteau, F. (1985) "Use of ultrasound scanning in swine for detection of pregnancy and some pathological conditions", Theriogenology 26, 267-278.

Cartee, R.E., Powe, T.A. and Ayer, R.L. (1985) "Ultrasonographic detection of pregnancy in sows", Modern Vet. Practice 66, 23-26.

Choi, H.S., Kiesenhofer, E., Gantner, H., Hois, J. and Bamberg, E. (1987) "Pregancy diagnosis in sows by estimation of oestrogens in blood, urine or faeces", Anim. Reprod. Sci. 15, 209-216.

Inaba, T., Nakazima, Y., Matsui, N. and Imori, T. (1983) "Early pregnancy diagnosis in sows by ultrasonic linear-electronic scanning", Theriogenology 20, 97-101.

Knight, J.W., Bazer, F.W., Thatcher, W.W., Franke, D.E. and Wallace, H.D. (1977) "Conceptus development in intact and unilaterally hysterectomized-ovariectomized gilts: interrelations among hormonal status, placental development, fetal fluids and fetal growth", J. Anim. Sci. 44, 620-637.

Madec, F., Martinat-Botté, F., Forgerit, Y., Le Denmat, M. and Vaudelet, J.C. (1988) "Utilisation de l'échotomographie en élevage porcin. Premiers essais de codification de quelques cas physiopathologiques concernant les organes génito-urinaires des truies", Rec. Méd. Vét. 164, 127-133.

Martinat-Botté, F., Botero, O, and Bariteau, F. (1985) "L'échotomographie: un diagnostic précoce de gestation chez la truie", Journées Recherch. Porc. en France 17, 155-162.

Martinat-Botté, F., Bariteau, F., Lepercq, M., Forgerit, Y. and Terqui, M. (1988) "L'échographie d'ultrasons, outil de diagnostic de gestation chez la truie", Rec. Méd. Vét. 164, 119-126.

Saba, N. and Hattersley, J.P. (1981) "Direct estimation of oestrone sulphate in serum for a rapid pregnancy diagnosis test", J. Reprod. Fertil. 62, 87-92.

van der Steen, H.A.M. (1985) "Economic evaluation of pregnancy diagnosis in pigs", Rapport no.4 Agricultural University Wageningen, The Netherlands.

Taverne, M.A.M., Oving, L., van Lieshout, M. and Willemse, A.H. (1985) "Pregnancy diagnosis in pigs: a field study comparing linear-array real-time ultrasound scanning and amplitude depth analysis" The Veterinary Quaterly 7, 271-276.

ACCURACY OF PREGNANCY DIAGNOSIS IN DOGS BY MEANS OF LINEAR-ARRAY ULTRASOUND SCANNING.

M.A.M. Taverne and H.A. van Oord
Department of Herd Health and Reproduction
State University of Utrecht
Yalelaan 7
3584 CL Utrecht
The Netherlands

INTRODUCTION

In the clinical assistance of canine reproduction, ultrasonography has been used for the diagnosis of pregnancy (Bondestam et al. 1983, 1984; Tainturier. 1984; Taverne et al. 1985; Shille et al. 1985), the prediction of littersize (Bondestam et al. 1984; Toal et al. 1986), the diagnosis of uterine pathology (Poffenbarger and Feeney 1986; Komarek 1986), the assessment of fetal viability in cases of threatened abortion, prolonged gestation and prior to caesarean section (Shille et al. 1985) and for the detection of ovulation (Inaba 1984).

The accuracies of detection of pregnancy and nonpregnancy by means of two-dimensional ultrasound scanning have been compared with those obtained by using abdominal palpation (Taverne et al. 1985; Toal et al. 1986). From these studies, performed on relatively small numbers of bitches, it was concluded that ultrasonography is slightly better than the hands of an experienced palpator. However, the major advantage of ultrasound over palpation is the more precise information on the developmental stage and viability of the fetuses.

Since we evaluated the results of transabdominal linear-array scannings performed in 1983, the procedure for dogs presented at the clinic for pregnancy diagnosis is as follows: each bitch is first investigated by abdominal palpation. Animals found to be pregnant are not further investigated unless the owner asks for an ultrasound scan. Animals in which the palpation is negative or doubtful (because the bitch was to fat or resisted the examination) are investigated by ultrasonography on the same day.

This paper reports on the results of these ultrasound examinations obtained from a group of 438 dogs during the period 1984-1987.

MATERIALS AND METHODS

The bitches used in this study were of 40 different breeds. The day of the first mating or insemination was always taken as day 0 of pregnancy. Special attention was payed to asking the owners when and how many matings had taken place. The relative frequency of the days after first mating on which the dogs were presented to the clinic are given in figures 1 and 2. The majority of the animals were between 25 and 35 days

M. M. Taverne and A. H. Willemse (eds.), Diagnostic Ultrasound and Animal Reproduction, 105–110.

of gestation. This type of selection was partly the result of instructions to owners to present their animals for abdominal palpation around day 30. At this time the local swellings of the uterus at the implantation sites are easier to detect, even when the actual age of the conceptus is a few days shorter (Concannon et al. 1983).

Ultrasound scans were performed by an experienced operator. During scanning the bitch was usually in standing position. If long hair or the size of the dog interfered with correct placement of the transducer, the bitch was laid on her side. In almost every case the midventral abdominal wall, between the two rows of teats, was shaved from the umbilicus to the os pubis. Owners did not object to this procedure.

Two types of ultrasound scanners were used: the XL-scanner of Diagnostic Sonar (Livingstone, Scotland) provided with a 3.0 or 5.0 MHz linear-array transducers and the scanner type 400 (Pie Medical, Maastricht, the Netherlands) with a 5.0 MHz linear-array transducer. After application of coupling gel to both the shaved area of the abdominal wall and the transducer the latter was always orientated initially along the longitudinal axis of the dog. Recognition of pregnancy was based on echographic pictures described earlier by Taverne et al (1985); depending on the stage of gestation the most relevant structures include: the yolk sac, the placenta and amniotic membrane and the embryos or fetuses themselves. No special attention was paid to the estimation of litter size although we informed the owners if we had seen only one, two, three or more than three fetuses.

Confirmation of the diagnosis was obtained by asking the owner (by telephone or letter) if the dog had whelped or not; date of whelping and littersize were also checked. In the few cases that no whelping followed a positive diagnosis we queried as to any discharge from the vulva after the ultrasound scan had been performed.

Knowing the outcome of (non-)pregnancies, the results were arranged as follows: the number of correct positive(a), incorrect positive(b), correct negative(c) and incorrect negative diagnoses(d). From these numbers sensitivity (a/a+d), specificity (c/b+c), positive predictive value (a/a+b) and negative predictive value (c/c+d) were calculated.

RESULTS

As a result of the selection of our patients, it could be expected that the majority (n=241) of all dogs (n=438) which we investigated did not whelp. Four of these animals were diagnosed as pregnant between 25-30 days after first mating. (Fig. 1). From these 4 bitches two were also diagnosed to be pregnant by abdominal palpation on the day of scanning; in 3 of them only 1-2 implantation chambers were seen during scanning and two of the four bitches showed discharge from the vulva at term.

From the 197 bitches which finally whelped, 11 were diagnosed as being not-pregnant between 27 and 32 after first mating. (figure 2). In table 1 the information on these 11 cases is summarized: 7 of the bitches delivered only 1 or 2 pups, 6 had been mated at least twice and 6 dogs delivered after the 65th day since the first mating.

On the basis of these results the following parameters were calculated:

Sensitivity (186/197): 94.5%
Specificity (237/241): 98.4%
Positive predictive value (186/190): 97.9%
Negative predictive value (237/248): 95.6%

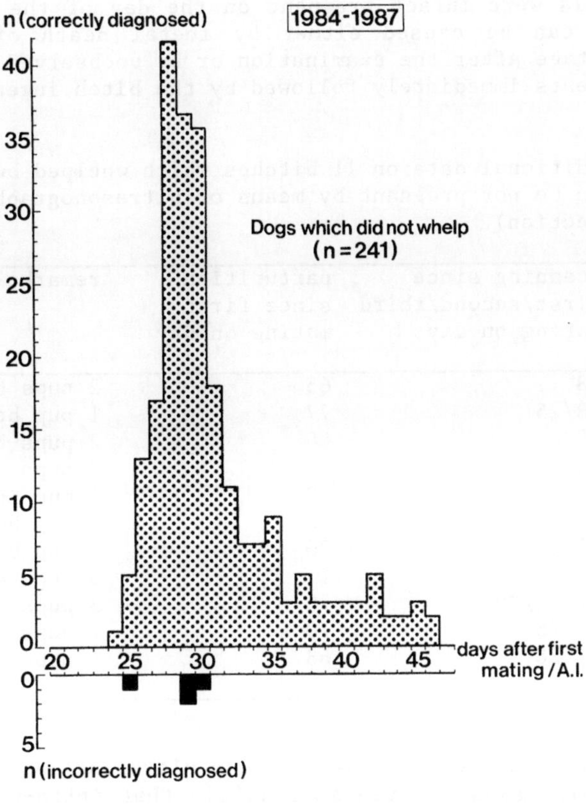

Figure 1. The number of bitches presented for pregnancy diagnosis between days 24 and 46 after their first mating and which finally did not whelp. (hatched area: correct negative diagnoses; black area: incorrect positive diagnoses).

DISCUSSION

The results of our study show that real-time ultrasonography is an accurate method for the diagnosis of pregnancy in bitches; they also illustrate that even in the hands of an experienced investigator, false diagnoses cannot be completely avoided. Compared to our first study (Taverne et al. 1985) in which no selection took place before dogs were presented for a scan, the present data contained more equal numbers of whelping and non whelping dogs; this enables a better comparison of the specificity and sensitivity of this diagnostic method.

In contrast to the studies (using small groups of patients) reported by Bondestam et al (1984) and Toal et al (1986) in which no false positive diagnoses were made, our results show that in 4 cases a positive diagnosis before day 30 was not followed by whelping. As in our first report (Taverne et al. 1985), there is evidence which suggests that these animals were infact pregnant on the day of the examination. The discrepancy can be caused either by foetal death of the entire litter at some stage after the examination or by unobserved expulsion of the uterine contents immediately followed by the bitch ingesting them.

TABLE 1. Additional data on 11 bitches which whelped but which were diagnosed to be not pregnant by means of ultrasonography (s.c.: caesarean section).

dog no.	scanning since first/second/third mating on day	parturition since first mating on day	remarks
1	28	65	2 pups born
2	28/25	77	1 pup born
3	28	66	2 pups born by s.c.
4	28/27	60	7 pups born by s.c.
5	30	62	1 pup born
6	27	62	1 pup born
7	28/25	70	1 pup born by s.c.
8	32	?	3 pups born
9	31/30/29	70	3 pups born
10	29/28	70	2 pups born
11	29/28	68	4 pups born

Indeed, only 1 or 2 fetuses were seen during ultrasonography in 3 of the 4 bitches and this makes it even more likely that intra-uterine foetal death or unobserved expulsion were responsible for the apparent false positive diagnoses. Assuming that these 4 dogs were indeed pregnant, our data indicate that only 2% of all the pregnancies (197+4) are lost once they have reached a stage between 25 and 45 days after first mating.

There are several explanations for the 11 false negative diagnoses in our study. Firstly, when only a small number of fetuses is present there will be a greater chance of overlooking them; indeed in 7 of the 11 cases only one or two pups were born at term. Secondly image quality may play an important role: adipose tissue in the abdominal wall or omentum and the presence of a lot of gas in the gastrointestinal tract sometimes make it very difficult to recognize any morphological details during scanning; when this is the case small litters once again will be easily missed. It remains to be established to what extend a period of

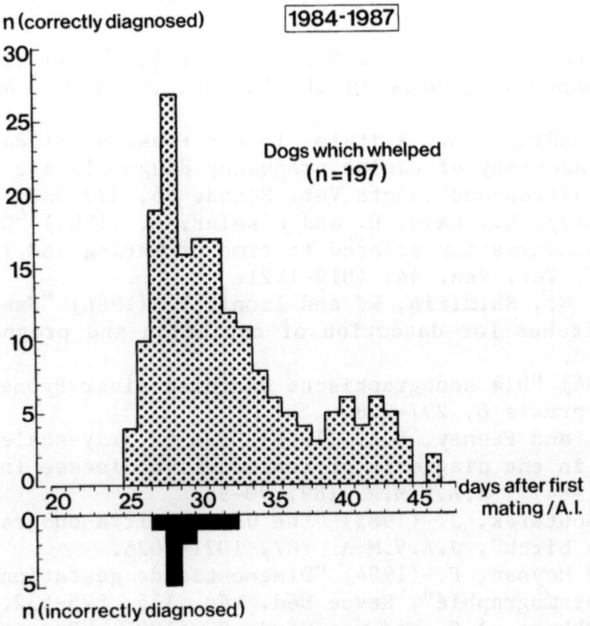

Figure 2. The number of bitches presented for pregnancy diagnosis between days 24 and 26 after their first mating and which finally whelped. (hatched area: correct positive diagnoses; black area: incorrect negative diagnoses).

starvation before scanning would lower the incidence of false negative diagnoses. Thirdly, there is no certainty about the exact age of the conceptus when the day of first mating is used for dating the pregnancy (Concannon et al. 1983). As a consequence, dogs may be scanned at a stage in which morphology of the pregnant uterus does not yet allow clear recognition of the uterine contents during transabdominal scanning. In our first report (Taverne et al. 1985) we provided evidence that this was the case in a few of our false negative diagnoses. In at least two dogs in the present study (cases 2 and 7 in Table 1) this might have been the case as well. In order to avoid such possibilities, the owners should always be asked exactly when and how many times the bitch has been mated. Fertilization may even take place several days after the last mating; so when several matings have taken place and a negative diagnosis is made rather early after the first mating the owner should be asked to present the dog for pregnancy diagnosis again one week later. In conclusion, transabdominal ultrasonography provides a highly accurate, additional diagnostic tool for the recognition of pregnancy and nonpregnancy in the dog; however, even when performed by an experienced investigator some mistakes (especially false negative diagnoses) can still be expected to be made, especially when animals with a small number of fetuses are involved.

REFERENCES

Bondestam, S., Alitalo, I. and Kärkkäinen, M. (1983) "Real-time ultrasound pregnancy diagnosis in the bitch", J. Small. Anim. Pract. 24, 145-151.

Bondestam, S., Kärkkäinen, M., Alitalo, I. and Foss, M. (1984) "Evaluating the accuracy of canine pregnancy diagnosis and litter size using real-time ultrasound", Acta Vet. Scand. 25, 327-332.

Concannon, P., Whaley, S., Lein, D. and Wissler, R. (1983) "Canine gestation length: variation related to time of mating and fertile life of sperm", Am. J. Vet. Res. 44, 1819-1821.

Inaba, T., Matsui, N., Shimizin, R. and Imoni, T. (1986) "Use of echography in bitches for detection of ovulation and pregnancy", Vet. Rec. 115, 276-277.

Komarek, J.V. (1986) "Die sonographische Diagnose einer Pyometra beim Hund", Kleintierpraxis 6, 297-298.

Poffenbarger, E.M. and Feeney, D.A. (1986) "Use of gray-scale ultrasonography in the diagnosis of reproductive disease in the bitch: 18 cases (1981-1984)", J.A.V.M.A. 189, 90-95.

Shille, V.M. and Gontarek, J. (1985) "The use of ultrasonography for pregnancy in the bitch", J.A.V.M.A. 187, 1021-1025.

Tainturier, D. and Moysan, F. (1984) "Diagnostic de gestation chez la chienne par echotomographié", Revue Méd. Vét. 135, 525-532.

Taverne, M.A.M., Okkens, A.C. and van Oord, R. (1985) "Pregnancy diagnosis in the dog: a comparison between abdominal palpation and linear-array real-time echography", The Veterinary Quarterly 7, 249-255.

Toal, R.L., Walker, M.A. and Henry, G.A. (1986) "A comparison of real-time ultrasound, palpation and radiographic in pregnancy detection and litter size determination in the bitch", Veterinary Radiology 27, 102-108.

DIAGNOSTIC ULTRASOUND IN OTHER AREAS OF VETERINARY MEDICINE

K.J. Dik, D.V.M., Ph.D., Professor of Radiology, Chairman,
Department of Radiology, Veterinary Faculty
Yalelaan 10
3584 CM Utrecht-de Uithof
The Netherlands

Introduction

Radiography provides excellent, detailed images of skeletal disorders, however the visualization of soft tissues is relatively poor.

Diagnostic ultrasound is effective especially for soft tissue imaging, not only to visualize pregnancy, ovarian - and uterine morphology and pathology, but also for many other parts of the body, such as the thoracic cavity including the heart, the abdominal cavity including liver, spleen, kidney, urinary bladder and prostate gland, bloodvessels like the aorta, iliac arteries, umbilical vessels, jugular vene and carotid artery, limb disorders like injuries to tendons, tendon sheaths, joints and muscles and miscellaneous lesions such as a soft tissue abscess, haematoma, cyst, tumor, or fistulous tract.

In contrary to radiography a sonogram (or scan) represents a 2-dimensional cross-section, thus avoiding superimposition and allowing accurate depth evaluation, but resulting in more difficult anatomical orientation.

Equipment

Obstretic-transrectal scanning of large animals is usually performed with a 3.0 - 5.0 MHz linear-array scanner.

Basically these scanners can be used in applications other than pregnancy however if superficial as well as deeper structures should be imaged in both large and small animals the equipment should be imaged in both large and small animals the equipment should be very vertasile;

- higher and lower soundwave frequencies must be available including a 3 MHz scanner to penetrate 20-24 cm deep tissue, a 5 MHz scanner providing a higher depth resolution but penetrating only 10-12 cm tissue and 7,5 (10) MHz for examining superficial structures and to ensure maximum resolution.
- A linear-array scanhead, ideal for transrectal examination, requires a long contact surface and is inadequate to scan other body areas such as the limbs or thorax ("shadowing" of the ribs).

M. M. Taverne and A. H. Willemse (eds.), Diagnostic Ultrasound and Animal Reproduction, 111–114.

- The ideal equipment to examine large and small animals and superficial as well as deeper tissues is a (mechanical) real-time sector scanner offering multifrequencycapabilities, with a (inline) scanhead design resulting in a small contact surface (7-9 mm^2) necessary to examine thin, contoured areas such as tendons less than 2 cm thick, as well as to scan irregular sufaces such as between the costal margins.

Furthermore the equipment should include a scanhead offset device to scan very superficial structures, M-mode display for quantitative evaluation of the heart, a polaroid or multiformat camera to obtain a permanent record of the scans and a videorecorder for permanent recording of dynamic events such as cardiac function.

Of course these instruments are more expensive ($ 50.000-150.000) than linear-array systems only used for obstretic imaging ($ 10.000 - 20.000).

Applications

1. Thorax

Diagnostic ultrasound is a particularly valuable imaging modality to characterize pleural effusion (clear or composite fluid, fibrin strands, adhesions, loculation, gas, tumor growth), including ultrasound-guided diagnostic or therapeutic thoracocentesis.

In addition pulmonary abscesses or lung consolidation extending to the surface of the lung will be visualized, although radiography is superior to diagnose intrapulmonary disease if not obscured by superimposing pleural fluid.

Ultrasound, as a non-invasive alternative for radiographic angiocardiography, is also capable to evaluate cardiac diseases such as pericardial effusion, congenital heart defects, valvular abnormalities (thickening, insufficiency, stenosis), atrial and/or ventricular enlargement as well as abnormal contractility and thickness of the myocardium.

Additional M-mode recording enables accurate measurement of the contractility, systolic and diastolic chamber size and cardiac wall thickness, as well as the valvular excursions.

Additional contrast echocardiography by jugular injection of saline boluses mixed with air, (or other CO_2 or H_2O_2 containing compounds) demonstrates interventricular or interatrial shunts.

2. Abdomen

Diagnostic ultrasound is also a valuable imaging technique to determine and characterize abdominal fluid, including ultrasoundguided abdominocentesis.

If abdominal fluid is not present (plain) abdominal radiography usually demonstrates size, shape and position of abdominal organs such as the liver, spleen, kidney, urinary bladder and prostate gland.

Diagnostic ultrasound also demonstrates the texture of these organs, thus providing valuable additional information to differentiate between abscessation, haematoma, cyst or tumor. Diagnostic ultrasound also visualizes (radiolucent) renal calculi, concrements in the urinary bladder and cholelithiasis and may be used for guided diagnostic biopsy procedures. Gas within the stomach and intestine completely reflects the sound beam, therefore sonographic examination will not allow assessment of intestinal diseases.

In small animals diagnostic ultrasound usually is performed in addition to the radiographic examination. The abdomen of (adult) large animals usually is too thick to obtain good-quality (plain) radiographs, in these animals diagnostic ultrasound has become the imaging modality of choice for examination of abdominal diseases.

3. Limb

Diagnostic ultrasound to evaluate limb disorders is frequently used in horses, particularly to examine tendon and tendon sheath injuries, less frequently to determine joint lesions and seldomly to evaluate muscular abnormalities.

For the diagnosis of tendon injuries the clinical examination (gait evaluation, inspection and palpation) is unreliable. Lameness and/or swelling may be absent or minimal, or masked by therapeutic measurements and differentiation between a peritendinous or tendinous swelling or effusion of a tendon sheath by palpation alone is inaccurate. The imaging of tendon injuries by plain radiography is very poor.

Diagnostic ultrasound enables precise evaluation of the location, extent and texture of the lesions. Follow-up serial examinations with an interval of 8-10 weeks enables monitoring of the healing progress thus allowing a more accurate prediction as to when the horse can withstand athletic challenge.

The usual technique to visualize joint lesions in addition to the clinical examination is plain radiography providing excellent detailed images of bones, however the radiographic imaging of soft tissue articular components such as the joint capsule, cavity and articular cartilage is very poor.

These components can be assessed by contrast radiography (arthrography) or arthroscopy, however these techniques are invasive, irritating and painful. Diagnostic ultrasound, although not yet frequently used to examine joints, may be a potentially attractive non-invasive screening technique to evaluate such lesions.

Diagnostic ultrasound is also a valuable, but seldomly used, imaging technique for the examination of muscle lesions such as fibrotic myopathy, foreign bodies, tears, haematoma, abscessation or tumor.

4. Miscellaneous

Diagnostic ultrasound is also a valuable technique to examine non-specific soft tissue swellings, such as abscess formation including ultrasound-guided puncture and pus removal, or to evaluate fistulous tracts and suspected foreign bodies, particularly foreign bodies not detectable on radiographs. In contrary to radiography ultrasound is also capable to determine precisely the depth and size of the object.

Normally diagnostic ultrasound is not used to examine skeletal disorders, due to the inability of the beam to penetrate bone.

However if radiographic examination is not possible, ultrasound may be used to evaluate bone surfaces, for instance to detect fractures (demonstrated by disruption or "step" defect of the bone contour), or roughening of the bone surface due to periosteal new bone formation.

In summary;

multiple imaging techniques, like endoscpoy, contrastradiography, xeroradiography, computed tomography, nuclear imaging, magnetic resonance imaging and ultrasound, are available to evaluate soft tissues.

Diagnostic ultrasound is a non-invasive, safe, relatively inexpensive imaging modality and therefore probably the most attractive diagnostic tool in veterinary practice to examine all kinds of soft tissue lesions.

Recommended Literature:

1. Vet. Clin. North Am. (Equine Pract.), Diagnostic Ultrasound, 2 (1), 1986 guest editor: Rantanen, R.W.
2. Vet. Clin. North Am. (Small Animal Pract.), Diagnostic Ultrasound, 15 (6), 1985 guest editor: Herring, D.S.

DIAGNOSTIC ULTRASOUND AND HUMAN REPRODUCTION

J.W. Wladimiroff, M.D., PhD
Department of Obstetrics & Gynaecology
Acedemic Hospital Rotterdam-Dijkzigt, Rotterdam
Erasmus University Rotterdam, The Netherlands

Diagnostic ultrasound plays a central role in present day of obstetric care. Embryonic development can be established as early as 5 weeks of gestation using transvaginal transducers. Fetal cardic activity can be observed at 6 weeks and fetal movements at 7-8 weeks. At about 10 weeks embryonic development is more or less completed. As from 14-15 weeks of gestation, reasonably detailed information can be obtained about fetal anatomy. At that time, ultrasound studies have clearly demonstrated the presence of an extensive number of dynamic properties in the fetus such as eye movements, swallowing, stomach and urinary bladder filling and emptying, breathing movements. Combined morphologic and dynamic studies of the fetus by ultrasound have lead to the possibility of detecting a large number of structural anomalies as early as 16-17 weeks of gestation. During the second half of pregnancy diagnostic ultrasound has demonstrated its value in the diagnosis of intrauterine growth retardation. Serial measurements on fetal head and upper abdominal circumference will provide information of fetal growth. Growth retardation usually is the result of poor placental perfusion i.e. chronic fetal hypoxia. Sometimes, however, growth retardation is associated with serious structural anomalies. Recently, the introduction of combined real-time and pulsed Doppler equipment has lead to studies on uterine and fetal blood flow. Reproducable changes in uterine artery and arcuate artery flow velocity waveforms as well as various fetal arteries such as umbilical artery, aortic and internal carotid artery have been collected during fetal hypoxia before growth retardation sets in. Diagnostic ultrasound also plays a very important supportive role with respect to the performance of various invasive procedures: chorionic villus sampling, amniocentesis and cordocentesis (umbilical cord punture).

1. THE FIRST HALF OF PREGNANCY (UP TO 20 WEEKS).

1.1 Chorionic villus sampling.

During chorionic villus sampling some chorionic villus tissue is collected under ultrasound monitoring for direct karyotyping, sex determination and diagnosis of some hereditory metabolic diseases. Sampling can be done via the trans-abdominal or trans-vaginal route. This sampling procedure is carried out between 9-11 weeks of gestation. However, lately our centre is increasingly using the abdominal approach after 11 weeks, that is during the second and even early third trimester of pregnancy (Pijpers et al, 1988). The advantage of both early and late chorionic villus sampling is that the result of karytyping will be available within 72 hours. This in contrast to amniocentesis, which is

M. M. Taverne and A. H. Willemse (eds.), Diagnostic Ultrasound and Animal Reproduction, 115–120.

carried out at 16 weeks with the result available 2 to 2½ weeks later. An early test with a quick result is of great psychological advantage to those parents who are at risk of an infant with congenital abnormality. If a lethal abnormality has been diagnosed before 13-14 weeks, pregnancy termination can be carried out by means of a suction curettage.

1.2 Early detection of fetal structural abnormalities.
This kind of diagnostic procedure will particularly be requested by those pregnant women who are at risk of a fetal structural abnormality. The risk factor can vary between 3-5% up to even 25-50%, depending on the pattern of inheritance. Proper genetic counselling with respect to the degree of risk and possibilities of prenatal diagnosis is essential. Ultrasound scanning for fetal structural abnormalities will usually be performed between 16-22 weeks of gestation. A new area on which attention is being focussed is the area of fetal cardiac abnormalities. The incidence of these abnormalities is 8 o/oo. A considerable number of cardiac abnormalities can now be diagnosed as early as 17-18 weeks of gestation. There are indications that the so-called colour-coded Doppler in which the presence and direction of flow can be recognized by colour will further improve the early detection of cardiovascular abnormalities.

1.3 Fetal movements.
A systematic qualitative description and classification of fetal movements during the first half of pregnancy using ultrasound has recently been published (de Vries et al, 1985). Different kinds of movements such as stretching, retroflexion, twisting, breathing and yawning have been described. Of interest is that in the presence of maternal inslulin dependent diabetes most movements up to 13 weeks of gestation show a developmental delay of 1-1½ week compared to the normal obstetric population. The significance of this observation is not yet clear (Visser et al, 1985).

2. ULTRASOUND DURING THE SECOND HALF OF PREGNANCY (ABOVE 20 WEEKS)

During the second half of pregnancy ultrasonic scanning is very much foccussed on the assesment of fetal growth and placental location. The early diagnosis of fetal growth retardation is now being studied using uterine, arcuate and fetal blood flow velocity waveforms. Also here the diagnosis and treatment of fetal structural and functional abnormalities should be mentioned. Invasive techniques such as cordocentesis under ultrasound monotoring has lead to new inroads of fetal diagnosis and treatment. The possibilities and use of intrauterine chirurgical under ultrasound monitoring in the presence of an abnormal fetus is currently an issue under discussion. More basic ultrasound studies are focussed on fetal behavioural states.

2.1 Early detection of fetal growth retardation.
Following the introduction of fetal blood flow velocity measurements using combined real-time and Doppler ultrasound equipment an increasing

number of reports points at the characteristic changes in the blood flowvelocity waveform, originating from the umbilical artery, the fetal descending aorta and fetal internal carotid artery with respect to fetal growth retardation as result of impaired uteroplacental perfusion (Campbell et al, 1983; Reuwer et al, 1984; Wladimiroff et al, 1986). Of clinical importance is the observation that changes in flow velocity waveforms in the umbilical artery and fetal descending aorta characterized by reduction in end-diagnostic flow velocity can already be documented several weeks prior to ultrasound measurement of fetal growth retardation itself (Fig. 1).

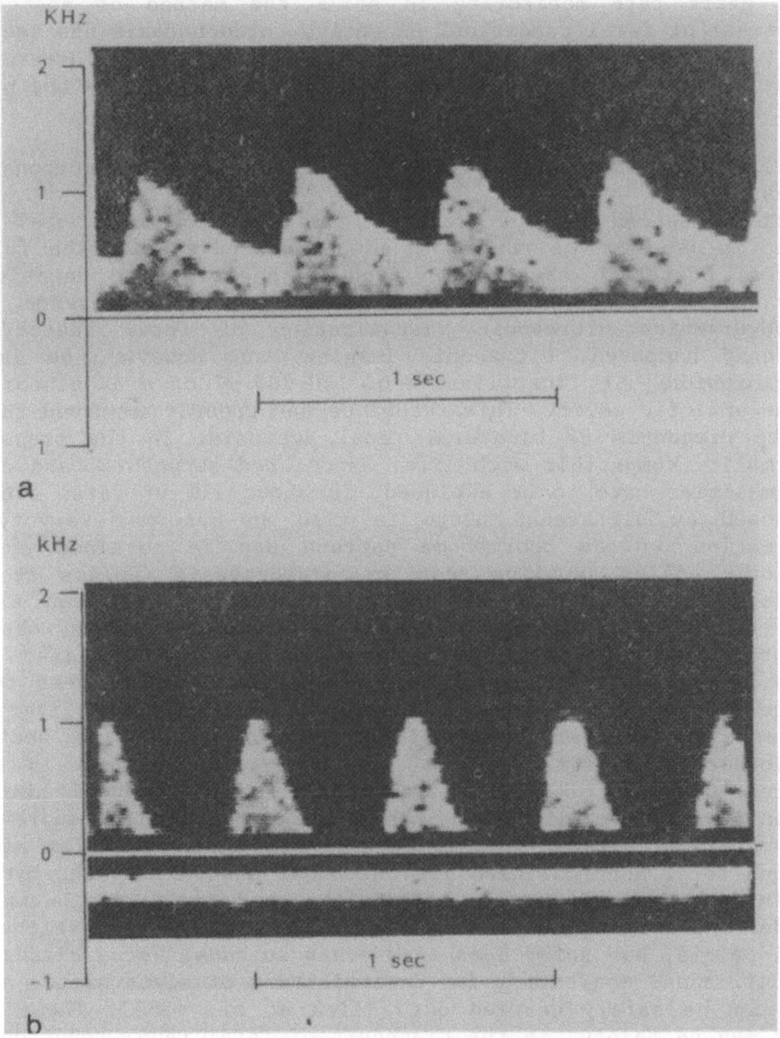

Figure 1.
Blood flow velocity waveform in the umbilical artery:
a. normal pattern with forward flow throughout the cardiac cycle;
b. abnormal pattern with absent end-diastolic flow indicating raised placental resistance due to impaired utero-placental perfusion.

Blood flow velocity waveforms have also been recorded on the maternal side of the placenta i.e. the arcuate arteries. A relation has been suggested between impaired end-diastolic flow velocity in these maternal vessels at 20-24 weeks and maternal hypertension at a later state. Studies of the sensitivity and specficity of these flow velocity waveforms in the early detection of intrauterine growth retardation are being carried out at various centres. A proper answer should be obtained from these studies before implementing these techniques in daily obstetric care.

2.2 Assessment of fetal condition.
Fetal heart rate monitoring is still the method of choice for the assessment of fetal condition. Recently, cordocentesis has been reported as a technique for obtaining fetal blood before labour. Determination of the acid-base status of the fetus can confirm or refute the presence of fetal hypoxia.

2.3 Diagnosis and treatment of fetal structural and functional abnormalities.
During the second half of pregnancy the detection of a fetal structural defect is usually the result of an ultrasound examination following an abnormal obstetric observation such as positive and negative discrepancy, haemorrage and premature labour. In the presence of severe oligohydramnios ultrasonic visualization of fetal anatomy will be seriously hampered. Ultrasonic imaging can, however, be improved by amnioinfusion, i.e. instillation of 150-200 ml of a 5% glucose solution in the amniotic cavity. This technique has greatly improved the accuracy of the diagnosis of bilateral renal agenesis. In the presence of an abnormality compatible with life, associated structural and chromosomal abnormalities have to be excluded. In about 11% of fetal abnormalities diagnosed by ultrasound there is also an abnormal karyotype. Rapid information on the chormosome pattern can be obtained by the late chorionic villus sampling test or cordocentesis (Daffos et al, 1985; Nicolaides, 1986). When an abnormal karyotype has been excluded, a consultation with the perinatal team consisting of an obstetrician, neonatologist, pediatric surgeon and clinical geneticist will take place. Attention will be focussed on I) the possible need for intrauterine medical or surgical treatment; II) premature delivery and III) the mode of delivery (vaginal delivery or caesarean section). The possibilities of fetal medical and surgical treatment is still very limited. One form of medical treatment is transplacental administration of antirarrhytmic drugs in the presence of fetal supraventricular tachycardia of atrial flutter. Surgical treatment is sofar aimed at two serious fetal abnormalities i.e. urethral obstruction and hydrocephaly. The introduction of a drain under ultrasound monitoring between urinary bladder and amniotic fluid compartment in order to relieve the dilatated renal pelvis, has sofar been a moderate success. Fetal bladder needling and ultrasound monitoring for determination of electrolytes and osmolarity can be safely carried out (Glick et al, 1985). These biomedical tests may be helpful in the diagnosis of fetal renal function. Shunting of dilatated cerebral ventricles in the presence of hydrocephaly has

been abanded due to the very poor results (Clewell et al, 1982). Moreover, it is impossible to properly assess brain function in the presence of hydrocephaly. The only form of invasive therapy which is of clinical significance is fetal blood transfusion via the umbilical vein in the presence of severe Rhesus disease.

2.4 Fetal behavioural states.

Simultaneous recording of fetal body and eye movements as well as fetal heart rate using ultrasound, has resulted in the recognition of fetal behavioural states in late pregnancy (Nijhuis et al, 1986). Quiet sleep states characterized by occasional body movements, absent eye movements and a heart rate with a small band-width should be distinguished from active sleep states whereby often fetal body and eye movements are present and the fetal heart rate pattern displays a wide oscillation pattern. This is of practical significance in that a healthy fetus may have a flat heart rate pattern which after sometime will become reactive. This should avoid the unnecessary diagnosis of fetal compromise (Nijhuis et al, 1986).

REFERENCES

Campbell, S., Diaz-Recasens, J., Griffin, D.R. et al (1983) "New Doppler technique for assessing uteroplacental blood flow", Lancet I, 675-677.

Clewell, W.H., Johnson, M.L., Meier, P.R. et al (1985) "A surgical approach to the treatment of fetal hydrocephalus", New Eng. J. Med. 306, 1320-1325.

Daffos, F., Capella-Pavlovsky, M. and Forester, F. (1985) "Fetal blood sampling during pregnancy with use of a needle guided by ultrasound: A study of 606 consecutive cases", Am. J. Obstet. Gynaecol. 153, 655-660.

Glick, P.L., Harrison, M.R., Golbus, M.S. et al (1985) "Management of the fetus with congenital hydronephrosis. II. Prognostic criteria and selection for treatment", J. Pediat. Surg. 20, 367-387.

Nicolaides, K.H., Rodeck, C.H., Soothill, A.W. and Campbell, S. (1986) "Ultrasound guided sampling of umbilical cord and placental blood to assess fetal well-being", Lancet I, 1065-1067.

Nijhuis, J.G. (1986) "Foetaal gedrag: hoeding van de foetus in nieuw perspectief", Ned. Tijdschr. Geneesk. 130, 1481-1485.

Pijpers, L., Jahoda, M.G.J. and Wladimiroff, J.W. et al (1988) "Transabdominal chorionic villus biopsy in 2nd and 3rd trimester pregnancies", Brit. Med. J. in press.

Reuwer, P.J.H.M., Bruinse, H.W., Stoutenbeek, Ph. and Haspels, A.A. (1984) "Doppler assessment of the fetoplacental circulation in normal and growth-retarded fetuses", Europ. J. Obstet. Gynaecol. Reprod. 18, 199-205.

Tonge, H.M., Wladimiroff, J.W., Noordam, M.J. and Kooten, C. van (1986) "Blood flow velocity waveforms in the descending aorta of the human fetus in the third trimester of pregnancy: comparison between normal and growth-retarded fetuses", Obstet. & Gynaecol. 67, 851-855.

Visser, G.H.A., Bekedam. D.J., Mulder, E.J.H. and Ballegooije, E. van (1985) "Delayed emergence of fetal behaviour in type I diabetic women", Early Hum. Dev. 12, 167-172.
de Vries, J.I.P., Visser, G.H.A. and Prechtl, H.F.R. (1985) "The emergence of fetal behaviour II. Quantitative aspects", Early Hum. Dev. 12 99-120.
Wladimiroff, J.W., Tonge, H.M. and Stewart, P.A. (1986) "Doppler ultrasound assessment of cerebal blood flow in the human fetus", Brit. J. Obstet. Gynaecol. 63, 471-475.

INDEX OF SUBJECTS

Current Topics in Veterinary Medicine and Animal Science

Recent publications

1984

26. Manipulation of Growth in Farm Animals, edited by J.F. Roche and D. O'Callaghan. ISBN 0–89838–617–8
27. Latent Herpes Virus Infections in Veterinary Medicine, edited by G. Wittmann, R.M. Gaskell and H.-J. Rziha. ISBN 0–89838–622–5
28. Grassland Beef Production, edited by W. Holmes. ISBN 0–89838–650–0
29. Recent Advances in Virus Diagnosis, edited by M.S. McNulty and J.B. McFerran. ISBN 0–89838–674–8
30. The Male in Farm Animal Reproduction, edited by M. Courot. ISBN 0–89838–682–9

1985

31. Endocrine Causes of Seasonal and Lactational Anestrus in Farm Animals, edited by F. Ellendorff and F. Elsaesser. ISBN 0–89838–738–8
32. Brucella Melitensis, edited by J.M. Verger and M. Plommet. ISBN 0–89838–742–6

1986

33. Diagnosis of Mycotoxicoses, edited by J.L. Richard and J.R. Thurston. ISBN 0–89838–751–5
34. Embryonic Mortality in Farm Animals, edited by J.M. Sreenan and M.G. Diskin. ISBN 0–89838–772–8
35. Social Space for Domestic Animals, edited by R. Zayan. ISBN 0–89838–773–6
36. The Present State of Leptospirosis Diagnosis and Control, edited by W.A. Ellis and T.W.A. Little. ISBN 0–89838–777–9
37. Acute Virus Infections of Poultry, edited by J.B. McFerran and M.S. McNulty. ISBN 0–89838–809–0

1987

38. Evaluation and Control of Meat Quality in Pigs, edited by P.V. Tarrant, G. Eikelenboom and G. Monin. ISBN 0–89838–854–6
39. Follicular Growth and Ovulation Rate in Farm Animals, edited by J.F. Roche and D. O'Callaghan. ISBN 0–89838–855–4
40. Cattle Housing Systems, Lameness and Behaviour, edited by H.K. Wierenga and D.J. Peterse. ISBN 0–89838–862–7
41. Physiological and Pharmacological Aspects of the Reticulo-rumen, edited by L.A.A. Ooms, A.D. Degryse and A.S.J.P.A.M. van Miert. ISBN 0–89838–878–3
42. Biology of Stress in Farm Animals: An Integrative Approach, edited by P.R. Wiepkema and P.W.M. van Adrichem. ISBN 0–89838–895–3
43. Helminth Zoonoses, edited by S. Geerts, V. Kumar and J. Brandt. ISBN 0–89838–896–1
44. Energy Metabolism in Farm Animals: Effects of Housing, Stress and Disease, edited by M.W.A. Verstegen and A.M. Henken. ISBN 0–89838–974–7
45. Summer Mastitis, edited by G. Thomas, H.J. Over, U. Vecht and P. Nansen. ISBN 0–89838–982–8

1988

46. Modelling of Livestock Production Systems, edited by S. Korver and J.A.M. van Arendonk. ISBN 0–89838–373–0
47. Increasing Small Ruminant Productivity in Semi-arid Areas, edited by E.F. Thomson and F.S. Thomson. ISBN 0–89838–386–2
48. The Management and Health of Farmed Deer, edited by H.W. Reid. ISBN 0–89838–408–7

1989

49. Vaccination and Control of Aujeszky's Disease, edited by J.T. van Oirschot. ISBN 0–7923–0184–6

MIX
Papier aus verantwortungsvollen Quellen
Paper from responsible sources
FSC® C105338

If you have any concerns about our products,
you can contact us on
ProductSafety@springernature.com

In case Publisher is established outside the EU,
the EU authorized representative is:
Springer Nature Customer Service Center GmbH
Europaplatz 3, 69115 Heidelberg, Germany

Printed by Libri Plureos GmbH
in Hamburg, Germany